T0320415

Sustainable Energy Pathways to Net Zero

Sustainable Energy Pathways to Net Zero addresses realistic pathways between now and 2050 to reach net zero: a steady state concentration of greenhouse gases in the atmosphere. It discusses solar and wind energy, hydrogen, energy storage, and electric vehicles, all of which are becoming an increasingly cost-efficient part of these pathways. Hydrogen, the price of electricity, the integration of different sources of electricity into the electrical grid, and the electric vehicle charging networks are crucial in balancing supply and demand. This book describes how these different energy factors fit together. It emphasizes the fact that the intersection of these technologies is where the most profound advances can occur.

Features:

- Emphasizes the importance of demand management for electricity and reducing cost.
- Explains the economics of reaching net zero emissions and the role of innovation and public policies, among others.
- Discusses the cost and efficiency of solar and wind power, electric vehicles, and storage technologies, and describes the benefits of battery swapping.
- Focuses on the role of integration of different sources of electricity into the electrical grid as an important part of the pathway.
- Recommends research and development on electrochemical processes to remove carbon from the ocean.
- Written in a simple language for a general audience and understandable for a global market.
- Is Open Access to encourage global use.

This book is a great resource for government and industry professionals involved in energy production and management, as well as academics and students in science and engineering interested in the pathways to sustainable development.

Sustainable Energy Pathways to Net Zero

Larry E. Erickson and Gary Brase

CRC Press is an imprint of the
Taylor & Francis Group, an informa business

Designed cover image: © dee karen, Shutterstock

First edition published 2025
by CRC Press
2385 NW Executive Center Drive, Suite 320, Boca Raton FL 33431

and by CRC Press
4 Park Square, Milton Park, Abingdon, Oxon, OX14 4RN

CRC Press is an imprint of Taylor & Francis Group, LLC

ISBN: 978-1-032-49941-3 (hbk)
ISBN: 978-1-032-49947-5 (pbk)
ISBN: 978-1-003-39615-4 (ebk)

DOI: 10.1201/9781003396154

Typeset in Times
by Apex CoVantage, LLC

Contents

Preface

We have a goal of reaching net zero emissions by 2050. That is, we have a goal of producing no more greenhouse gasses than the amount being absorbed by the ecosystem. This net zero goal is not only a logically necessary target for countering global climate change; it is included in the Paris Agreement on climate change, and there are thus many efforts in progress already to reduce emissions. Each country that is part of the Paris Agreement is developing a plan to reduce emissions, and there are annual United Nations Conferences of the Parties (COP) to review progress. This is a process associated with the United Nations Framework Convention on Climate Change (UNFCCC). A wide range of people and organizations are involved in research, development, and implementation in support of reaching the net zero goal; enough of a range that it can be difficult to look at the overall picture of all the things happening.

This book addresses a number of the major topics that are key to achieving a transition to net zero by 2050, with many of these topics being processes of transitions themselves. Very often, the situation is that there are ways in which the world has been doing things—in terms of energy production, delivery, and storage, in terms of transportation, in housing, and in agriculture—and these ways have to change. Across all these areas, there are transitions either happening now or on the near horizon. There are many pathways to reach net zero, however, and some pathways are better than others. There are differences in quality of life, economics, and citizen preferences associated with the choices that are being made and the pathways those choices lead to.

One of the very important developments is progress in the generation of electricity using solar and wind energy. The goal of the UNFCCC is to triple the capacity of solar- and wind-generated electricity by 2030. These renewable energy technologies lead to a need for improved demand management practices and technologies that can make sure that energy is reliably available to consumers and that all of the renewable electricity that is produced is used effectively.

The next set of transitions for a net zero emissions future has to do with the things that electricity goes into: cars, trucks, other types of vehicles, heating and cooling of homes, lighting, and even the global agricultural industry.

Finally, there are transitions in terms of decisions and behaviors. Net zero emissions have to be workable in terms of economics. Policies need to be consistent with those economics and the need to align incentives so that the world moves toward -rather than further away from—a net zero future. And those economic and policy considerations eventually lead to people making decisions every day. We, therefore, need to understand how people make decisions and how the environment can be structured to support environmentally responsible behaviors.

Part of people making good decisions is education; people need to know what the realistic options are, what should be done, and what needs to be done in order to reach net zero. This book is an effort to develop content that is easy to read for those who want to grow in their understanding of the transitions toward net zero. This book is intended for educational programs, for adult education, and for those in leadership positions. It is also written for general readers.

It can be used with our two earlier books: *Reducing Greenhouse Gas Emissions and Improving Air Quality* (Erickson and Brase, 2020) and *Solar Powered Charging Infrastructure for Electric Vehicles* (Erickson et al., 2017). We have found that, as we have continued to delve into these topics, there is an ever-growing circle of interrelated topics that extend our understanding of where the world currently stands on this issue. We hope that you will also find this to be the case.

References

Erickson, L.E. and Brase, G. 2020. Reducing Greenhouse Gas Emissions and Improving Air Quality: Two Interrelated Global Challenges. CRC Press, Boca Raton, FL.

Erickson, L.E., Robinson, J., Brase, G. and Cutsor, J., Eds. 2017. Solar Powered Charging Infrastructure for Electric Vehicles: A Sustainable Development. CRC Press, Boca Raton, FL.

Acknowledgments

There are many activities related to the topics in this book that occur at Kansas State University, in Kansas, the USA, and globally that contribute to our understanding of the Paris Agreement on Climate Change, the importance of net zero greenhouse gas emissions, and the pathways to reduce emissions. Our interactions with faculty, staff, and students have been very beneficial, as has our participation in professional and environmental organizations.

We specifically thank the United Nations for permission to include the printed version of the address "A Moment of Truth" on June 5, 2024, by Antonio Guterres, Secretary General of the United Nations, who spoke at the American Museum of Natural History in New York.

At the end of the day, though, there is always a specific core of people who helped us in this particular endeavor, along with those who perennially help us in our lives. We, therefore, express our special thanks to Irma Britton, Danita Deters, Sheree Walsh, Sandra Brase, and Laurel Erickson.

Larry E. Erickson
Gary Brase

About the Authors

Larry E. Erickson is a professor of chemical engineering at Kansas State University. In May 2024, he completed 60 years of service at K-State. His research interests include biochemical and environmental engineering, air quality, and sustainable development. His degrees are in chemical engineering from K-State.

Gary Brase is a professor in the Department of Psychological Sciences at Kansas State University. He studies complex human decision-making using social, cognitive, and evolutionary theories. His research includes work on topics such as medical decisions, decisions about sustainability issues, relationship and fertility decision-making, personality and mating decisions, and reasoning about social rules. Dr. Brase has over 75 journal and book chapter publications and over 100 research presentations. He has been at K-State since 2007.

1

Introduction

1.1 Net Zero

As I begin writing this book in the summer of 2023, there are major wildfires in Canada. The extent and duration of these fires have created air pollution that drifted south and into a large portion of the United States. Whereas normal air quality is about 50 ppm (parts per million) or less of pollutants, on June 28, the reported values of the Air Quality Index were 216 in Chicago, IL, 255 in Grand Rapids, MI, and 306 in Detroit, MI (Bilefsky, 2023).

Air pollution from wildfires is not just a nuisance; weather and climate disasters such as these actually kill people every year. The total reported deaths from the largest 357 weather and climate events from 1980 to June 2023 in the U.S. (the ones that cost at least a billion dollars in damage) are 15,957 deaths or about 45 people per event (NCEI, 2023). The estimated total cost of these 357 events was $2.545 trillion.

Most people who experience poor air quality do not die, of course, but air pollution negatively impacts the health of many more people in less severe ways. The 2023 wildfires in Canada impacted air quality and health from Minnesota to the eastern states in the U.S. as well as many people in Canada. Air quality impacts associated with climate change are one of the important reasons for taking action to help stop climate change.

There were some short-term options—at least in the case of the Canadian wildfires—to avoid these situations. The forests in these parts of Canada are near water in many cases. Irrigation could have been used to supply water to the trees during very dry periods to improve the health of the vegetation and thus reduce the chances of fire. There would be a cost to supplying this water during very dry conditions, but a small fraction of the much greater costs associated with the wildfires.

Short-term options are fine when they are available and there is no time for anything else. There is a long-term situation, though, that we need to address. Why did the Canadian forests get so unusually dry? Basically, it was hotter and drier than usual; the weather patterns were shifting. The latest prediction from the World Meteorological Organization (WMO, 2023) is that the annual mean global near-surface temperature for each year between 2023 and 2027 is

DOI: 10.1201/9781003396154-1

predicted to be between 1.1C and 1.8C higher than the average over the years 1850–1900. This is global climate change.

One very bad wildfire season, or even all the recent bad wildfire seasons, is serious enough, but that is not the only effect of global climate change. Climate change increases both the frequency and severity of all sorts of weather events: fires, floods, droughts, hurricanes, and more. Each of these leads to direct damages and injuries or deaths. Longer term, they degrade the environment, lead to poorer water quality and air quality, and increase the spread of some pests and diseases. This can include increases in the range of disease-spreading organisms (e.g., ticks). So, there are negative consequences from climate change in terms of respiratory issues, heart diseases, and pathogen-related illnesses, in addition to problems like starvation, displacement, increased crime, and overall poor mental health (EPA, 2023).

The long-term situation, which needs a long-term solution, is global climate change. Global climate change is a lot more than worse wildfires; it is more frequent and more extreme flooding, drought, hurricanes, and more. The long-term solution to climate change is to stop putting so many pollutants into the atmosphere that are creating and maintaining those changes. Easily said but not easy to do. First of all, we will never completely stop emitting pollution into the atmosphere, nor do we need to. Living on this planet tends to generate some emissions, and our planet has mechanisms to deal with those emissions. Secondly, though, we do need to stop producing quantities of emissions that far outstrip those planetary coping mechanisms. We need to get to "net zero," where the emissions being produced are no more than those emissions being removed. (We could even spend some time in a net negative situation to help deal with past emissions, but let's focus on getting to zero for now.)

The essence of net zero as a goal is embodied by an idea that is over a hundred years old, known as the "first law of holes." The first law of holes is: *If you find yourself in a hole, stop digging* (Wikipedia, 2023). There are a number of metaphorical questions that come from this idea: How deep is the hole now? How fast are we still digging? Great questions that are probably useful for various things. The first thing to do, though, is to stop digging.

The goal of net zero is to reach a point where the concentration of greenhouse gasses in the atmosphere stops increasing, which means the emissions of greenhouse gases are equal to (or less than) the rate of removal. The efforts to reduce emissions and transition to net zero have been going on for many years, and much progress has been reported (Kerry and McCarthy, 2021). In fact, the goal of reaching net zero greenhouse gas emissions by 2050 has become the current ambitious but possible goal for sustainable development globally. Accordingly, many people and organizations have devoted their attention and efforts to planning, conducting research, developing processes, implementing changes, and introducing new technologies to advance sustainable development in many countries based on this goal.

Global efforts can be made to meet all of the Sustainable Development Goals (UN, 2015; Erickson and Brase, 2020) and improve the quality of life for everyone everywhere. In addition to the Nationally Determined Contributions of countries toward meeting the goals of the Paris Agreement (UNFCCC, 2022; Erickson and Brase, 2020), there can be net zero commitments by cities, regions, businesses, financial institutions, and other organizations such as universities, churches, and societies (UN, 2022). There are many actions that are important in the implementation of these commitments, and many people are involved.

The recommendation of the high-level expert group of the United Nations (UN, 2022) to cities, etc., is:

> A net zero pledge should be made publicly by the leadership of the non-state actor and represent a fair share of the global climate mitigation effort. The pledge should contain interim targets (including targets for 2025, 2030 and 2035) and plans to reach net zero in line with IPCC or IEA net zero greenhouse gas emissions modelled pathways that limit warming to 1.5C with no or limited overshoot and with global emissions declining to at least 50% by 2030, reaching net zero by 2050 or sooner. Net zero must be sustained thereafter.

1.2 Energy Production

This book starts with the topics of increasing the use of solar and wind to generate electricity (Chapters 2 and 3) because of the expectation that this is the best pathway to reaching the goals of the Paris Agreement on Climate Change. Reducing the use of coal, petroleum, and natural gas is necessary to reach net zero emissions. There is sufficient wind and solar energy on the planet to replace all of the carbon that is being used for fuel. However, more sophisticated management is needed to balance supply and demand as the amount of wind and solar energy generation increases.

Wind and solar energy have become low-cost sources for generating electricity. Both wind and solar radiation are free and available globally. One of the advances that are likely to make wind and solar energy even more attractive is the production of hydrogen-powered by renewable electricity. Our ability to make this "green hydrogen" is quickly advancing, with several alternative methods of electrolysis of water (Ishaq et al., 2022). In order to reach net-zero carbon emissions, increases in wind and solar electric power generation are needed to advance toward a zero-carbon electrical grid and toward a supporting green hydrogen production network to have hydrogen available for dispatch to power fuel cells to produce electricity when needed.

The cost and efficiency of solar panels have enabled many people to install them on their homes and businesses to produce electricity that is beneficial economically and environmentally. The progress in improving solar panels is one of the more important developments along the path to net-zero emissions. Further developments are anticipated (IEA, 2022; IRENA, 2023).

1.3 Energy Demand Management

The use of different energy production methods, like solar and wind power, involves the use of different management systems of that energy. Sunlight and wind are natural and renewable resources with patterns of ebbs and flows. Because the fluctuating supply of solar and wind energy will not fit perfectly with the fluctuating demands of people, the management system is more complicated but nowhere near impossible. Chapter 4 is about what this "smart" electrical grid will look like.

Solar and wind energy management is not only about meeting the demands of consumers at any given time but also about using energy from solar and wind generation that is not immediately needed. When there is more energy produced than people demanding it at the moment, there are several management options. One option is to store that excess energy in batteries, in large storage systems by energy distributors, or even in all the batteries consumers have.

Demand management for electricity is an important concept. The progress in developing solar and wind technologies to produce electricity has been excellent, and renewable sources of electricity are the least expensive in many communities (IEA, 2022). This progress means, however, that we will soon have times when some solar and wind power systems are generating not only enough energy for all their regular customers but also excess energy to be put to use somehow.

Demand management can be used to make sure that there is always a market for the electricity that is generated by renewable sources (Erickson and Ma, 2021). When there is abundant energy, batteries can be charged, and hydrogen can be generated by electrolysis of water to provide ways to manage demand.

Prices for electricity can also be adjusted as part of the demand management process. Time-of-use prices enables customers to charge batteries when the cost is low. In fact, one basic way to generally manage demand is through pricing. For example, electric vehicles (EVs) can be charged at low prices in parking lots with solar-powered charging infrastructure. As electric vehicles (EVs) become increasingly common, their batteries can be essentially storage devices if they are charged while energy is abundant. Demand management prices (DMPs) are one tool that nudges consumers to make good use of the electricity generated by wind and solar sources. Introducing a lower DMP for charging batteries of electric vehicles that are parked during the day in parking lots equipped with solar panels and EV charging equipment basically shifts that energy demand to when the supply is higher. The DMP is a reduced price to provide an incentive for drivers to participate in the program and allow their EVs to be charged as needed to balance supply and demand.

At a more industrial level, hydrogen can be produced by electrolysis of water at low prices when excess supply exists in the electrical grid. At peak power demand, the price for power can be large enough to pay for more of the cost of electricity from wind and solar. Hydrogen that is produced by the electrolysis of water with renewable energy has many uses. Ammonia can be produced and used for fertilizer and as an energy source. Methane can be produced from hydrogen, water, and carbon dioxide. Ammonia, methane, and hydrogen can be stored. The production of hydrogen, ammonia, and methane can be increased as needed to make use of all of the electricity that is being produced (for example, during the spring and fall, when the demand for electricity for heating and cooling is lower).

Both of these uses of electricity are designed to be used as needed to make use of electricity generated in excess of the regular consumer demand at that particular time. What about the other side of balancing supply and demand? What about when there is greater demand for electricity than the supply from wind and solar generation? Some cynics express this by noting that "the sun doesn't always shine, and the wind doesn't always blow." The most direct way to manage this is simply to have supplementation from other energy production methods (e.g., fossil fuels or nuclear power). But there is also a major management role here for the rapidly developing area of storage methods for previously generated solar and wind energy. There are many opportunities to make use of hydrogen and ammonia as fuels to generate electricity when a dispatchable fuel is needed, as a fuel for ships and other uses. There are also numerous other developing technologies for storing energy in media, such as batteries and hydropower.

Chapter 5 focuses on maintaining and improving the reliability and resilience of our energy system as it transitions to mainly solar and wind energy. The quality of our electrical grid must be considered and managed as part of the efforts to move toward net zero greenhouse gas emissions. Our current electrical system is generally very reliable and resilient . . . except, of course, when it isn't. Winter storm Uri in February 2021 is the most expensive winter weather disaster in the United States. It produced widespread electrical outages and associated deaths, mostly in Texas (FERC, 2021). Analyses of this winter storm and associated power issues found a number of failures and generated many recommendations to prevent similar problems in the future. Some of these recommendations were about power systems and better demand management. Coincidentally, some of the power management strategies that come with renewable energy resources—battery storage of power, multiple power sources that are adaptively used, and more local generation of energy at the place of use—would also be powerful steps to improve the reliability and resilience issues in a place like the Texas energy grid.

1.4 Specific Energy Uses

In addition to the broad, structural aspects of generating energy from renewable sources and having an energy management network that is designed to effectively use that energy, there are a number of specific uses of energy that need to be renovated in order to help us reach a net zero status. Typically, the adjustments here are either to replace fossil fuel-based technologies or to be more efficient (i.e., use less) fossil fuels if they are still required. Some major areas we cover in this book include transportation (Chapter 6), housing (Chapter 7), and agriculture (Chapter 8).

When it comes to transportation and renewable energy, one of the obvious ongoing transitions is the adoption of electric vehicles (EVs). Many people have experienced the pleasure and benefits of electric vehicles (EVs), which are now common in many parts of the world. These authors have both been driving to work using EVs for electricity for many years now. For local transportation, EVs are great. Most people can plug an EV in at home to charge using a normal outlet. (Higher voltage charging can be installed, of course, but it is not necessary for most normal driving patterns.) Charging and maintenance costs are also exceptionally low for EVs. In the last 10 years, there has been significant progress in the development of EVs, particularly progress in batteries and infrastructure to support the electrification of transportation. The challenge ahead is to electrify all transportation and eliminate greenhouse gases associated with transportation (Kerry and McCarthy, 2021).

Another important task is to reach net zero where people live and work by using renewable energy for heating and cooling, heating water, and cooking (Kerry and McCarthy, 2021). Multiple aspects of our housing involve the use of energy: heating in the winter, cooling in the summer, and lighting all the time. These are most commonly powered by electricity from the power grid (see Chapters 4 and 5), but there are also a number of advances that already exist, are ongoing, or are coming soon to increase structural efficiency. These are changes we will be able to implement for our homes, workplaces, schools, public buildings, and commercial buildings.

Agriculture is another important topic in reaching net zero, and in fact, reducing greenhouse gas emissions is needed throughout the entire food supply system. On farms and ranches, the electrification of tractors and other equipment and the shift to using renewable nitrogen fertilizer are both very significant issues. Production of renewable electricity and carbon sequestration are also possible on the land used for agriculture. In principle, net zero ethanol can be produced for use as a liquid fuel once the transitions have been made to electric tractors and net zero fertilizer on a net zero farm.

Further along the food supply chain, food waste is a significant issue that should be managed better. Waste reduction through good management is important because using food for its intended purpose is much better than discarding it. Even the processing of unwanted food, though, can be improved. The waste stream that goes to a landfill can be better managed to reduce methane emissions associated with anaerobic microbial processes in the landfill (Erickson and Pidlisnyuk, 2021; FAO, 2021; UNEP and FAO, 2020).

As mentioned earlier, hydrogen is very important to the transition to green energy because ammonia can be produced using green hydrogen and nitrogen for fertilizer and for energy. Low-cost fertilizer can be produced from low-cost solar energy that is used for hydrogen production via electrolysis when the demand for electricity needs to be increased to balance supply and demand. Hydrogen can also be used to produce methane from captured carbon dioxide. Ammonia and methane produced from green hydrogen can then be stored and used as needed.

Multiple areas such as solar energy, wind energy, green hydrogen, energy storage, and electric vehicles are each becoming increasingly cost-efficient, and they are also all part of a pathway. Crucially, it is at the intersection of these technologies where some of the most profound advances can occur. There are manifold effects of combining solar, wind, and hydrogen energy, along with wider storage and usage options. The integration of different sources of electricity into the electrical grid, the electric vehicle charging network, and the production of hydrogen by electrolysis are all parts of the pathway to reach net zero. Hydrogen production by electrolysis of water can be fundamentally important to the balance of supply and demand for electricity. All electricity that is generated using wind and solar energy should be used productively. Whether it is used for charging electric vehicles, storage batteries, or the production of hydrogen, generated power must be managed such that there is always a demand for electricity at a low price that is attractive to some users.

In short, rapid changes in both the cost and efficiency of solar and wind power, electric vehicles, and storage technologies pose an array of challenges, but they pose even more opportunities for improvements in the environment and public health.

1.5 Humans in the Loop

People in engineering sometimes refer to "humans in the loop" as a way to describe the role of humans within the systems they are building. Whether it is an electrical system, computer system, infrastructure system, or something else, there is usually a human in the loop somewhere—often in the center of things because the system is designed for them to use.

Humans are a key factor in the systems that exist now, producing excessive greenhouse gas emissions and global climate change. Humans will also be a key factor in developing and running a new system that gets us to net zero emissions. Chapter 10 looks at the larger dynamics of the current energy system, the changes it is currently undergoing, and the changes still to come. This includes the economics of sustainable energy and the politics of sustainability.

Sometimes there are people who facilitate and push for change. The United Nations (UN, 2022) has a good set of recommendations for cities and other organizations to follow in their efforts to follow a proposed net zero pathway to their goal. Sometimes (or at the same time) there are people who oppose and create obstacles to change. It is important to understand the motives and incentive structures in society that produce these forces opposed to changes.

Chapter 12 takes a more psychologically based approach to understanding the role of humans in addressing global climate change and reducing emissions to net zero. We consider how people think about climate (the world and its resources around us) and the decisions they make as consumers within the world.

1.6 Conclusion

This book finishes with a chapter on the international aspects of getting to net zero. Greenhouse gas emissions do not stay within any international borders and global climate change happens throughout the entire globe. That means solutions to these problems need to be on a global scale; net zero emissions need to be the future of the world. Failing to coordinate and work internationally on climate change will lead to conflicts over dwindling non-renewable resources, climate refugees, and wars. Ironically, wars and other conflicts are also associated with even more greenhouse gas emissions, making precipitating factors even worse.

At this point, the reader should begin to understand that action is needed by individuals, families, schools, churches, universities, cities, regional and state governments, nations, businesses, etc. to reach net zero. Many organizations have developed climate action plans with goals that include net zero greenhouse gas emissions by 2050 (MARC, 2021; London, 2022; Sacramento, 2022). Many of these plans have goals with earlier dates. The City of London plans to reach net zero carbon emissions from its own operations by 2027. Sacramento County has a goal to achieve carbon neutrality by 2030. Kansas City has a goal to reach net zero for local government operations by 2030, followed by a goal of net zero carbon emissions associated with electricity generation by 2035 (MARC, 2021).

The Paris Agreement on climate change was adopted on December 12, 2015, by the parties of the United Nations Framework Convention on Climate Change (UNFCCC), and after ratification by the parties, it came into force on November 4, 2016 (Erickson and Brase, 2020; UNFCCC, 2016). One of the goals of this agreement is to keep the increase in global average temperature to less than 2C above the global average temperature during the pre-industrial period from 1850–1900. Many of the representatives of the countries want to keep the average global warming to less than 1.5C.

The Paris Agreement addresses a global commons issue: the concentrations of greenhouse gas concentrations in the atmosphere. Each country is taking action to reduce emissions in support of the goals of the Paris Agreement, and we can use the excellent work by Elenor Ostrom on methods to address common issues to make progress toward meeting the net zero goal.

The plans of the United States include reaching net-zero greenhouse gas emissions by no later than 2050 (Kerry and McCarthy, 2021). To reach that target, the plan sets a goal of 100% clean electricity by 2035. The U.S. is also participating in the global methane pledge to reduce global methane emissions by at least 30% by 2030.

Ultimately, the goal is to maintain the current standards of incredibly high reliability and resilience of energy for consumers at about the same price but move to renewable energy sources in order to stop generating the pollution creating climate change. There are a number of complex parts to figure out in order to get to this goal, but none of these parts are impossible. Many of the parts are either already worked out or are fairly easy to address. The economics of reaching net zero emissions depend on good ideas, innovation, public policies, human behavior, and the quality of many companies, governments, and organizations. Many people are already driving electric vehicles (EVs) because of the personal economic benefits. Others are riding in electric buses. The air quality in large cities has improved because of the electrification of transportation. There is a path to sustainable development and net zero emissions by 2050. This book provides a map of that pathway.

References

Bilefsky, D. 2023. What to Know about Canadian Wildfires and U.S. Air Quality. New York Times, June 29.

EPA. 2023. United States Environmental Protection Agency. Climate Change and Human Health, February 27. www.epa.gov/climateimpacts/climate-change-and-human-health

Erickson, L.E. and Brase, G. 2020. Reducing Greenhouse Gas Emissions and Improving Air Quality: Two Interrelated Global Challenges. CRC Press, Boca Raton, FL.

Erickson, L.E. and Ma, S. 2021. Solar-powered charging networks for electric vehicles. Energies 14: 966.

Erickson, L.E. and Pidlisnyuk, V., Eds. 2021. Phytotechnology with Biomass Production: Sustainable Management of Contaminated Sites. CRC Press, Boca Raton, FL.

FAO. 2021. Principles for Ecosystems Restoration to Guide the United Nations Decade 2021–2030. UN Food and Agriculture Organization.

FERC. 2021. The February 2021 Cold Weather Outages in Texas and the South Central United States. FERC—NERC—Regional Entity Staff Report, Federal Energy Regulatory Commission and North American Electric Reliability Corporation, November 2021.

IEA. 2022. Renewables 2022. International Energy Agency. iea.org/

IRENA. 2023. Innovation Landscape for Smart Electrification. International Renewable Energy Agency, Abu Dhabi. irena.org/

Ishaq, H., Dincer, I. and Crawford, C. 2022. A review on hydrogen production and utilization: Challenges and opportunities. International Journal of Hydrogen Energy 47: 26238–26264.

Kerry, J. and McCarthy, G. 2021. The Long-Term Strategy of the United States: Pathways to Net-Zero Greenhouse Gas Emissions by 2050. U.S. Department of State, Washington, DC.

London. 2022. Climate Action Strategy 2020–2027. City of London Corporation. cityoflondon.gov.uk/

MARC. 2021. Regional Climate Action Plan. Mid-America Regional Council. kcmetroclimateplan.org/

NCEI. 2023. Billion-Dollar Weather and Climate Disasters. National Centers for Environmental Information, June 29. ncei.noaa.gov/access/billions/

Sacramento. 2022. Sacramento County Climate Action Plan. Sacramento County, CA. planning.saccounty.gov/

UN. 2015. Transforming Our World: The 2030 Agenda for Sustainable Development. United Nations.

UN. 2022. Integrity Matters: Net Zero Commitments by Businesses, Financial Institutions, Cities, and Regions. United Nations.

UNEP and FAO. 2020. The UN Decade on Ecosystem Restoration 2021–2030. UN Environment Programme. unep.org/

UNFCCC. 2016. Paris Agreement. United Nations Framework Convention on Climate Change, United Nations.

UNFCCC. 2022. 2022 NDC Synthesis Report. United Nations Framework Convention on Climate Change. United Nations, October 26.

Wikipedia. 2023. Law of Holes, October 22. https://en.wikipedia.org/wiki/Law_of_holes

WMO. 2023. WMO Global Annual to Decadal Climate Update. World Meteorological Organization, Geneva. wmo.int/

2

Solar Energy

2.1 Introduction

Solar Energy is free. The sun shines, and radiant energy comes to the earth, which is used by plants and people. In many ways, solar energy is the foundation of not only a sustainable energy future but the foundation of life on earth. The first law of thermodynamics says that the energy of a closed system must remain constant—it can neither increase nor decrease without interference from outside. Although our planet is in some respects a closed system, it crucially gets a constant and large input of energy from the sun into that system. Plants convert that solar energy into food (chemical energy), allowing for most of life to exist.

It only makes sense that solar energy is one of the important sources of energy for the generation of electricity using photovoltaics (PV). The prices for solar PV installations are such that many solar panels are being installed on homes and other buildings to generate electricity. Parking lots have been covered with solar panels to provide shade for parked vehicles and to generate electricity that can be fed to the grid and be used to provide charging for EVs (Erickson et al., 2017).

Because of its present low cost and the expectation of lower costs in the future, solar-generated electricity is expected to become a very important source of energy in the future. The cost of electricity from utility-scale solar PV has decreased by 85% in the decade from 2010 to 2020 (IRENA, 2022). Renewables (solar and wind) are expected to grow to become very important sources of electricity in many countries by 2027 (IEA, 2022). Solar-generated electricity is also expected to increasingly be a source of energy to produce hydrogen via electrolysis of water. Between 2023 and 2030, renewables are expected to become the largest source of electricity in the world. The installed capacity of solar PV is expected to exceed that of coal in 2030 (IEA, 2022).

Of course, there are some issues to confront with solar energy. Solar radiation varies during the day and when clouds are present. The resulting fluctuations in solar energy need to be managed.

DOI: 10.1201/9781003396154-2

Battery storage helps to manage supply and demand at a system level when solar energy is a major source of electricity. When used as part of an electrical grid, other sources of electricity and battery storage can be used together to manage supply and meet demand.

2.2 Solar PV

There has been rapid growth in the installation of solar PV systems in many parts of the world, with the greatest installed capacity being in Asia (Allouhi et al., 2022; IEA, 2022). Allouhi et al. (2022) provide information on the path to reduced prices for solar panels and review several issues, such as dust management, panel temperature effects, and cleaning alternatives. At the end of 2021, the global installed capacity was 843 GW for solar PV (IRENA, 2022). While growth in solar capacity has been good, the amount added per year needs to continue to increase toward the goal of adding 800 GW of renewable power per year (IRENA, 2022). The goal for 2030 is to have over 11,000 GW of renewable power capacity globally (IRENA, 2023). There is a need to increase the percentage of renewable electricity and increase the production of hydrogen from renewables (IRENA, 2022).

The price of utility-scale electricity from solar PV was about $0.04/kWh in 2022 (IRENA, 2022). Because of the low prices for electricity, wind and solar installations have been leading in the amount of new installed capacity in 2020–2022, and this is expected to continue. In locations with excellent solar conditions, such as Qatar, United Arab Emirates, and Saudi Arabia, values below $0.02/kWh have been reported (IRENA, 2022).

Electricity from solar PV has no direct carbon emissions; it is produced when the sun is shining.

There are some carbon emissions from the manufacturing of solar panels: about 40–50 grams of CO2 per kilowatt-hour of electricity generated. There are also concerns about the mining and transportation of the materials used in building solar panels. Materials like silver, copper, indium, and tellurium sometimes come from problematic locations and conditions. We need to be mindful, however, that the perfect should not be an enemy of the good. If solar energy is being used instead of gas or coal energy, the manufacturing emissions are offset very quickly. Natural gas produces 117 lbs of CO2 per million British thermal units (MMBtu) during extraction and production. Oil produces 160 lbs of CO2 per MMBtu (Wigness, 2023). Additionally, fossil fuels (oil, gas, and coal) also are derived from problematic locations and conditions. The geopolitical issues due to fossil fuel dependence have led U.S. presidents to call repeatedly for energy independence over the last 60 years. Solar and wind power could finally get the U.S. to that independence. Furthermore, fossil fuels contribute *additional* carbon emissions when used as an energy source.

2.3 Location, Location, Location

The cost of electricity produced by solar panels on homes and other buildings varies because of local solar conditions, labor costs, and other factors, but it is often in the range of $0.06–0.13/kWh (DOE, 2021; Ramasamy et al., 2022). To the extent that solar panels are located at or near the place where the generated electricity will be used, these are efficiency gains because there are minimal transmission structure costs or transmission losses. There are often savings on electricity costs associated with adding solar panels to homes and businesses. The cost of solar-generated electricity using solar panels is the lowest for solar panels in fields compared to solar panels on buildings and highest for solar panels in parking lots because of the cost of the supporting structure. All of these three alternatives are expanding with new installations. There are many positive reports on economics and good return on investment for solar-generated electricity (Blok, 2023; IEA, 2022; Matasci, 2023; Nugent, 2022).

There are also positive externalities in some situations; for instance, solar panels in parking lots provide shade for vehicles and do not take up any additional land because that space is already being used for parking. Presently, 10–15% of the solar generation area that is in use is located in parking lots, and very large quantities of parking lot area remain to be used for solar PV (Nugent, 2022). There is sufficient area for new solar installations in most parts of the U.S. if all buildings and parking lots are considered.

Another option that has merit is to locate solar panels on lakes and reservoirs to reduce the evaporation of water (Farrar et al., 2022; Ravichandran and Panneerselvam, 2022). The efficiency of electricity generation is higher because the temperature of the solar panels is lower (Asmelash and Prakash, 2019; Dorenkamper et al., 2021). Because of wind and waves, as well as changes in water level, there are challenges in managing floating PV systems.

Another alternative is to use agrivoltaics to produce electricity with solar PV and reduce the solar radiation of the crop because of partial shading. Lettuce and strawberries are examples of plants that benefit from shading. One of the benefits of agrivoltaics is increased income per unit area of land (Klokov et al., 2023). Another is improved efficiency in the use of water for crop production. The total installed capacity of agrivoltaic systems is more than 2.8 GW (Klokov et al., 2023).

In the effort to reduce greenhouse gas emissions, solar-generated electricity ideally needs to increase to a point at which the electrical grid can operate without carbon emissions. In the U.S., the goal is to be able to have electrical grids with zero carbon emissions in 2035 (Kerry and McCarthy, 2021). To accomplish this, all electricity generated using wind and solar should be either used or stored at all times. The first priority for use is generally to serve the immediate electrical needs of residential and business customers because this is a time-sensitive need (people need that electricity now, not at some later time). A second priority should be short-term storage of energy to meet customer demand whenever solar and wind energy production is lower than demand. A third priority can be non-time-sensitive uses of energy.

A prime example of this final use for excess energy is using solar-generated electricity to produce green hydrogen via the electrolysis of water. One concept is for the electrolysis process to be variable in rate such that the price of electricity to produce green hydrogen is much lower, as it is always secondary to the primary customers. Because solar-generated electricity occurs when the sun shines, it is important to use prices to always have a demand for all of the electricity that is generated.

A similar concept of demand management is included in a recent publication where a parking lot covered by solar panels with electric vehicle (EV) charging stations has the capability to charge EV batteries at a reduced price when demand needs to be increased to balance supply (Erickson and Ma, 2021). Prices can be used for demand management to accomplish the goal of always having a demand for solar-generated electricity. Parking lots are an excellent location for solar PV because the solar panels provide shade, which is beneficial, and the land is used for multiple purposes (parking, EV charging, demand management).

After 2035, when there is sufficient renewable electricity and nuclear generation to supply all electricity without carbon emissions, there should be some secondary demands for electricity, such as the charging of batteries and the production of hydrogen that are accomplished at a lower price because they can be interrupted if the electricity needs to be used for a higher priority. There will need to be battery charging for several purposes including powering the grid when electricity is needed for that purpose. Battery charging can be accomplished when the supply of electricity is sufficient to meet this secondary demand. Commercial charging stations for EVs may operate at time-of-use prices to reduce demand at times of peak power use.

Another alternative that can work with solar-generated electricity is charging EV batteries separately from the actual vehicle (when the power is available) and then swapping out batteries in the actual vehicles. The concept of battery swapping is being implemented in China and some other countries because there is greater choice as to when batteries are charged (Cui et al., 2023). One of the important advantages of battery swapping is that it can be accomplished quickly compared to recharging a battery (Cui et al., 2023; Baldwin, 2023). Because of this, the range of the swappable battery can be less than that of a battery that needs to be recharged. For large trucks, the range of the swappable auxiliary battery can be selected to be appropriate for the distance between stops the driver prefers. For large trucks, the weight of the batteries is important. Drivers of large trucks that have more than one battery with at least one swappable battery have more choices in their energy management. For driving a long distance, there is also the concept of a swappable auxiliary battery that might be attached to the vehicle or towed behind it.

In the United States, battery swapping has started in California with a focus on locations where there is a demand for the service. Robots are used, and the time required to complete the swap is less than 10 minutes (Baldwin, 2023). One of the implications of a battery swapping system is that the batteries may be owned by the company that runs the battery swapping service. Customers would pay either by month or by usage for access to the batteries. On the consumer side of this system, it means that their EV would not actually include a battery *in situ*. Because the battery is a major part of the cost of an EV, an EV that is part of a battery-swapping system would be much less expensive to buy.

Finally, one common case to consider for energy needs is a large event where many come to participate (e.g., a sporting event or concert). With swappable battery opportunities, preparations can be made and charged batteries can be available for those who need them on the day of the event. Solar PV provides electricity when it can, and batteries are charged when electricity is available at an acceptable price. Some who attend the event may come with an auxiliary battery that they add when needed for traveling longer distances. With swappable batteries, there are two options because they can be charged or swapped.

References

Allouhi, A., Rehman, S., Buker, M.S. and Said, Z. 2022. Up-to-date literature review on solar PV systems: Technology progress, market status and R&D. Journal of Cleaner Production 362: 132339.

Asmelash, E. and Prakash, G. 2019. Future of Solar Photovoltaic. International Renewable Energy Agency, Abu Dhabi.

Baldwin, R. 2023. Why charge an EV when you can swap its battery? The Verge, July 27. theverge.com/

Blok, A. 2023. Solar parking lots are win-win energy idea. Why aren't they the norm. CNET, February 13. cnet.com/

Cui, D., Wang, Z., Liu, P., Wang, S., Dorrell, D.G., Li, X and Zhan, W. 2023. Operation optimization approaches of electric vehicle battery swapping and charging station: A literature review. Energy 263: 126095.

DOE. 2021. 2030 Solar Cost Targets. Solar Energy Technologies Office, Energy Efficiency and Renewable Energy. U.S. Department of Energy. energy.gov/eere/solar/

Dorenkamper, M., Wahed, A., Kumar, A., deJong, M. and Kroon, J. 2021. The cooling effect of floating PV in two different climate zones: A comparison of field test data from the Netherlands and Singapore. Solar Energy 219: 15–23.

Erickson, L.E. and Ma, S. 2021. Solar-powered charging networks for electric vehicles. Energies 14: 966.

Erickson, L.E., Robinson, J., Brase, G. and Cutsor, J., Eds. 2017. Solar Powered Charging Infrastructure for Electric Vehicles: A Sustainable Development. CRC Press, Boca Raton, FL.

Farrar, L.W., Bahaj, A.S., James, P., Anwar, A. and Amdar, N. 2022. Floating solar PV to reduce water evaporation in water stressed regions and power pumping: Case study Jordan. Energy Conservation and Management 260: 115598.

IEA. 2022. Renewables 2022: Analysis and Forecast to 2027. International Energy Agency. iea.org/

IRENA. 2022. World Energy Transitions Outlook. International Renewable Energy Agency, Abu Dhabi. irena.org/

IRENA. 2023. World Energy Transitions Outlook 2023. International Renewable Energy Agency, Abu Dhabi. irena.org/

Kerry, J. and McCarthy, G. 2021. The Long-Term Strategy of the United States: Pathways to Net-Zero Greenhouse Gas Emission by 2050.

Klokov, A.V., Loktionov, E.Y., Loktionov, Y.V., et al. 2023. A mini-review of current activities and future trends in agrivoltaics. Energies 16: 3009.

Matasci, S. 2023. Solar canopies: Bring solar panels to our parking lot. September 19. energysage.com/

Nugent, C. 2022. The overlooked solar power potential of America's parking lots. Time, December 8. time.com/

Ramasamy, V., Zuboy, J., O'Shaughnessy, E., et al. 2022. U.S. Solar Photovoltaic System and Energy Storage Cost Benchmarks, with Minimum Sustainable Price Analysis: Q1 2022. Technical Report NREL/TP-7A40–83586, National Renewable Energy Laboratory. nrel.gov/

Ravichandran, N. and Panneerselvam, B. 2022. Floating photovoltaic system for Indian artificial reservoirs—An effective approach to reduce evaporation and carbon emission. International Journal of Environmental Science and Technology 19: 7951–7968.

Wigness, S. 2023. What is the Carbon Footprint of Solar Panels? August 31. www.solar.com/learn/what-is-the-carbon-footprint-of-solar-panels/.

3

Wind Energy

3.1 Introduction

There has been great progress in the development of the science and technology for producing electricity using wind. Wind is free, just like the sun, and electricity can be generated inexpensively using wind turbines on land and on water. In 2022, the cost of electricity generated on land was about $29/MWh (2.9 cents/kWh) in the United States (Stehly and Duffy, 2022). This value varies from site to site because of the quality of the wind (Wiser and Bolinger, 2023). Electricity from wind is one of the least expensive sources of electricity. Because of the low cost, electricity from wind has expanded rapidly since 2000 (Erickson and Brase, 2020). At the end of 2022, the total installed capacity was 906 GW for electricity generated from wind globally (Hutchinson and Zhao, 2023). The wind capacity in the United States was about 140 GW (EIA, 2023). This compares to a global installed capacity of about 8000 GW, which includes 4436 GW using fossil fuels, 3026 GW for renewables, and 375 GW of nuclear (Statista, 2023). Because of the need to balance supply and demand, the 4436 GW includes a standby capacity that can be used when it is needed. All electricity generated by wind should be used to supply the needs of the grid, charge batteries, or produce hydrogen.

The Global Wind Energy Council (GWEC) estimates that 680 GW of new wind capacity will be installed during the five years from 2023–2027 (Hutchinson and Zhao, 2023). In Europe, there is a need to add capacity because of the impact of the invasion of Ukraine on energy security. In the U.S., the Inflation Reduction Act is expected to have positive impacts on adding renewable energy capacity during the next 10 years. The expectation is that wind generation capacity will pass 1000 GW (which is 1 TW) in 2023 and 2 TW by the end of 2030 (Hutchinson and Zhao, 2023). The generating capacities of both wind and solar energy are projected to grow significantly between now and 2050, not only because of the importance of decreasing greenhouse gas emissions but also because they are increasingly better options economically and politically. There is a need to increase the use of renewables in the grid as well as use more electricity from renewables to produce hydrogen using electrolysis of water.

DOI: 10.1201/9781003396154-3

3.2 Onshore Wind Energy

Most of the electricity that has been generated using wind has been accomplished on land and in China, Europe, and the USA, but many additional countries are starting to produce electricity using wind. Brazil and India are moving forward with plans to increase their generation of renewable electricity using wind (Hutchinson and Zhao, 2023). The state of the onshore wind industry has matured and expanded through the experiences of the past 20 years. It is now time to expand to other countries and move ahead with the increased growth that is needed to meet the goals of the Paris Agreement to reach net zero greenhouse gas emissions by 2050. There is significant progress in expanding wind-generated electricity in Africa.

Approximately nine GW of capacity has been identified in African countries as of 2023 (GWEC, 2023A). Angola, Egypt, Ethiopia, Ghana, Kenya, Morocco, Namibia, Nigeria, Senegal, South Africa, Tanzania, and Tunisia are among the countries with wind farms. Egypt has plans to increase its capacity to more than 25 GW using wind. The plan includes reaching 42% of electricity from renewables by 2030. Grid stability is better when there is a mix of solar and wind (GWEC, 2023A). There is a significant generation using hydro in Africa; wind and solar generation provide beneficial diversity. In 2023, 140 different wind projects were identified in Africa with an estimated total capacity of 86 GW if all are completed (GWEC, 2023A).

In the U.S., wind supplied 62% of electricity in 2022 in Iowa, 55% in South Dakota, and 47% in Kansas. For the entire U.S., 10% was supplied by wind in 2022, so many additional states have wind resources to have a majority of wind-supplied energy. At the end of 2022, 41 hybrid wind generation facilities that included storage and/or solar have been installed in the U.S. Wind and battery storage are the most common wind hybrid; however, solar and battery storage are being constructed as well (Wiser and Bolinger, 2023). The Inflation Reduction Act is expected to impact the future expansion of wind generation of electricity in the U.S. positively. There are several new incentives to encourage new installations (Wiser and Bolinger, 2023).

Other countries are similarly poised to have very large shares of wind power. There were a total of 22 countries with 10% or greater percentage of electricity from wind in 2022 (Wiser and Bolinger, 2023), and the country of Denmark supplied 57% of its electricity from wind.

3.3 Offshore Wind Energy

There are also many projects developing offshore wind energy as well. Why put wind turbines out in the ocean instead of on land? Offshore wind turbines are generally more efficient for several reasons: the winds are stronger, more consistent, and with less interference from geographical or human obstacles.

There is progress in developing offshore generation of electricity from wind in many parts of the world (Williams and Zhao, 2023). Fixed bottom systems are installed in shallow waters, whereas floating foundations are used in deep water. Larger wind systems may be used offshore because of transportation issues on land. The cost of electricity generated by offshore wind in 2022 was $73/MWh for fixed-bottom installations. The estimated cost of electricity from floating bottom systems is $133/MWh (Stehly and Duffy, 2022). The science and technology for offshore wind generation of electricity is advancing; with further development, a cost of $51/MWh is estimated for 2030 (Stehly and Duffy, 2022). As of 2023, the estimated installations over the next 10 years (2023–2032) are 380 GW of new offshore wind generation capacity (Williams and Zhao, 2023).

When this is added to the 64.3 GW that was in service at the end of 2022, 444.3 GW of offshore capacity will be available (Williams and Zhao, 2023). Good cooperation between governments and industry is beneficial to achieve this goal. Good working partnerships are very important in large projects.

In 2022, 60 MW of new offshore floating wind-generating capacity was added by Norway. The total capacity is now 171 MW in this region. The forecast is 10.9 GW of floating wind generation capacity by 2030 (Williams and Zhao, 2023).

3.4 Issues

There are social and community issues associated with the development of wind energy projects. Externalities must be considered, and zoning regulations must be followed. People may be concerned about wind farm noise or visual impact, as well as its effects on birds or other wildlife. For example, offshore wind installations are generally far enough from the coast that noise and visual issues are minimal. Good communication is needed with regard to project development, benefits to society, and impacts on the community. A good review of the literature on these issues is by Van Rensburg et al. (2020). The concept of a social license to operate is introduced as an aspect of developing relationships between the project leadership and members of the community. Addressing distributional justice issues is an important concern in many communities where negative externalities exist.

Public engagement and stakeholder participation in wind energy development has also been reviewed (Solman et al., 2021). Much of the literature addresses public participation in local wind projects. There has been significant activity by governments to develop incentives and regulations to address concerns associated with externalities. The transmission of electricity from wind farms to where it is used is another area where public engagement is important.

A final issue that is becoming common with wind power is curtailment. Curtailment is when the output of a power source is intentionally reduced below what it could have otherwise produced. This is a situation that already has arisen for large wind farms when very favorable wind conditions would allow the farm to produce more electricity than the local market has a use for. The curtailment of wind-generated electricity was 5.2 % in the U.S. in 2022 compared to the goal of using all of the electricity productively.

One of the benefits of a hybrid wind farm that includes batteries is that curtailment can be reduced by charging batteries with excess electricity. As the fraction of electricity generated from renewables increases, challenges related to the management of the grid increase, and greater efforts are needed to avoid curtailment (Wiser and Bolinger, 2023). Transmission of electricity to where it can be used is one aspect that must be considered because transmission lines can reach their capacity.

References

EIA. 2023. Short-Term Energy Outlook. U.S. Energy Information Administration. www.eia.gov/

Erickson, L.E. and Brase, G. 2020. Reducing Greenhouse Gas Emissions and Improving Air Quality: Two Interrelated Global Challenges. CRC Press, Boca Raton, FL.

GWEC. 2023A. The Status of Wind in Africa October 2023. Global Wind Energy Council, Brussels, Belgium. www.gwec.net/

Hutchinson, M. and Zhao, F. 2023. Global Wind Report 2023. Global Wind Energy Council, Brussels, Belgium. www.gwec.net/

Solman, H., Smits, M., van Vliet, B. and Bush, S. 2021. Co-production in the wind energy sector: A systematic review of public engagement beyond invited stakeholder participation. Energy Research and Social Science 72: 101876.

Statista. 2023. Global Installed Electricity Capacity 2021, by Source, Statista. statista.com/

Stehly, T. and Duffy, P. 2022. 2021 Cost of Wind Energy Review. National Renewable Energy Laboratory. www.nrel.gov/

Van Rensburg, T., Carr, N., Fitzpatrick, C., Curran, R. and Totterdell, C. 2020. Literature Review: Earning Local Support for Wind Energy Projects in Ireland. Sustainable Energy Authority of Ireland. www.seai.ir/

Williams, R. and Zhao, F. 2023. Global Offshore Wind Report 2023. Global Wind Energy Council, Brussels, Belgium. www.gwec.net/

Wiser, R. and Bolinger, M. 2023. Land-Based Wind Market Report: 2023 Edition. U.S. Department of Energy. doe.gov/

4

The Smart Electrical Grid

4.1 Introduction

What is a "smart grid"? Although there are some varying definitions, a good and concise one to start with is that a smart grid is an electrical network that includes two-way communication of information to intelligently integrate the actions of all its components in order to have efficient transmission, distribution, and use of electricity. There are several good definitions of the term smart grid, reviewed by Dorji et al. (2023).

Having a smart electrical grid is crucial for reaching net zero greenhouse gas emissions because of the challenges related to integrating demand management, variation in renewable energy generation, hydrogen production, battery storage, distributed generation, and electric vehicle charging. The quality of the communication infrastructure is very important for smart grid operations (Suhaimy et al., 2022). There is a chapter on the smart grid in our earlier book (Erickson and Brase, 2020) which contains useful information. A very good recent article on the smart grid has been published (Kabeyi and Olanrewaju, 2023).

We do not yet have fully operational smart grids in most of the world, but we are well on the way to it. There has been good progress in the implementation of the smart grid in the United States, with about 73% of residential customers having advanced metering infrastructure (AMI) in 2022. The AMI smart meter is an essential part of the smart grid. You can think of the AMI as the foundational "eyes and ears" of a smart grid that has to be in place for any good communication and integration.

4.2 Features of a Smart Grid

There are many important features of a smart grid that make use of developments in science and technology, including automatic monitoring and control, the internet of things, and artificial intelligence (Dorji et al., 2023; Escobar et al., 2021). The monitoring aspect requires many sensors (e.g., AMIs) that collect data that is used to make decisions that improve reliability, efficiency,

DOI: 10.1201/9781003396154-4

and security. Communication systems and managed prices for electricity allow customers to make decisions that improve the efficiency of distribution and use of electricity. With a smart grid and time-of-use prices, electric vehicle owners can charge their EVs at times when prices are low, using metering systems that collect data continuously for billing and grid management. Battery storage systems can be integrated into the grid and used with real-time data to improve grid management, reliability, and stability.

The smart grid includes many areas where electricity is distributed to customers, the larger grid where utilities are connected to other utilities, and the power pool where electricity may be purchased and sold through independent system operators. There are many substations associated with the grid that collect data that is used for grid monitoring and control.

4.3 Demand Management

At the heart of a smart grid is effective and continuous demand management. To understand why, let's first consider the existing (non-smart) electrical grid. Management is largely (ignoring many nuances) about increasing or decreasing the supply of electricity to match momentary demand. For example, an oil or gas power plant increases or decreases its output, which is something that can be controlled by the plant operators. Now consider what happens with renewable power generation: the supply of electricity is also increasing or decreasing over time and not entirely in synch with the increases or decreases in demand. There are now two different fluctuations that need to be managed—both demand and supply. Getting supply and demand to match at any given time can theoretically involve four possible actions: increasing supply, decreasing supply, increasing demand, and decreasing demand. There is never a need to do all of these at once, of course. In fact, some options are preferable to others. If the energy demands are being more than met by renewable energy generation, one could "decrease supply" by curtailing the energy generation, but that is essentially refusing "free" energy. A much better option in such cases is to find a way to use the additional energy to increase demand. We have already mentioned several options for essentially increasing demand when there is available energy: putting it into batteries, or producing green hydrogen. Furthermore, excess energy stored at one point in time can then be used at a later time to fulfill the demand that outstrips the renewable energy generation at that later time (i.e., increase supply). Thus, it is important that a smart grid includes features that can be used for demand management issues because renewable energy produced by wind and solar energy depends on nature.

Another way to increase demand to be in balance with energy generation is to adjust the prices for energy when supply is high. Energy distribution companies are already familiar with this, as they confront price changes in the energy market all the time. The only addition here is a broader adjustment to the prices of electricity at the consumer level. Changing prices for electricity and communicating the availability of the reduced price to users can effectively increase demand. For example, it is possible to set up a simple system that charges batteries in electric vehicles when demand needs to be increased by having a reduced demand management price (DMP), a special discounted price when energy supplies are high, and demand needs to be increased to match it. This opportunity is communicated to parking lot managers who control EV charging, who implement the change in demand (Erickson and Ma, 2021).

Solar-powered charging infrastructure for EVs is available in many parking lots (Erickson et al., 2017; Erickson and Ma, 2021). A smart grid can include features that enable the implementation of a DMP to charge EVs that are parked in parking spaces with solar panels and charging equipment that is connected to the smart grid. EV customers can park their EV, connect it to the charging station, and communicate their desire to receive electricity at the DMP. The grid operator can supply electricity at this reduced price when demand needs to be increased, and the DMP is the best price that is available. One of the benefits of a smart grid is the ability to automate this process.

Energy demand can also be increased by charging batteries that are used in battery swap operations. One important advantage of battery swap technology to charge EVs is that the batteries can be charged when prices for electricity are low and demand needs to be increased. The time required for the swap is also small. When there is a major event, batteries can be charged in advance of the event at times when demand needs to be increased in the grid. On the day of the event, batteries can be swapped to serve the needs of those who drive EVs to the event. Battery swap technology is good for EVs that are in service most of the time because of the short time needed for the battery swap (Feng and Lu, 2021; Korcer et al., 2022). There are many battery-swapping stations in China (Wu, 2022).

The smart grid has many features that may be used for the integration of EVs into the grid. Sultan et al. (2022) review this literature, which addresses issues that affect demand management. For instance, although, in most cases, the flow of electricity is from the grid to batteries, EVs can also supply electricity to a grid. EVs can also supply electricity to locations that are not connected to the grid, such as campsites.

Blockchain technology has been developed to keep track of data on the sale of electricity to EV customers through the smart grid. With variations in prices to help manage demand, there is a need to have blockchain technology to keep track of all of the transactions securely and with transparency for all who need to be informed (Kabeyi and Olanrewaju, 2023; Hasankhani et al., 2021).

The smart grid is beneficial because of new developments that are beneficial to both customers and producers (Escobar et al., 2021). The ability to add renewable power and manage it effectively to reduce greenhouse gas emissions and move toward net zero emissions has benefits for everyone. Cooperative efforts among regulators, electricity rate-setting commissions, public utilities, and citizens to find ways to use the smart grid beneficially are encouraged. Smart cities can find ways to improve the quality of life by developing advances that are beneficial for citizens and society through smart grid applications.

4.4 Communication

Communication systems have improved greatly since 1990, and there are many recent developments within smart grids as well. With the addition of modern communication options, customers can benefit from opportunities to help with demand management by using electricity when demand needs to be increased. In some cases, these opportunities can be automated and implemented by agreement.

Two good reviews on communication technologies for the smart grid provide useful information (Abrahamsen et al., 2021; Suhaimy et al., 2022). There has been good progress in the implementation of the smart grid, with about 99% of installations having advanced metering infrastructure (AMI) in Norway and 71% of European consumers with AMI as of 2021.

Globally, the estimate is about 800 million smart meters as of 2021 (Abrahamsen et al., 2021). After the installation of smart meters, features of the smart grid, such as time of use rates for electricity, can be implemented. The smart meters can report electrical use hourly. Fault detection systems can report problems and identify locations.

Improvements in reliability, resilience, and stability are goals that are being met in many installations. Monitoring of power quality at many locations allows frequency, voltage, and waveform to be maintained within desired limits. The management of distributed power generation is an important feature of the smart grid. Good monitoring and control are important in integrating distributed power into the grid.

There are many networks and systems used for communication. There are premise networks (HAN = home area network), neighborhood area networks (NAN), and wide area networks (WAN), and these can be wired or wireless (Abrahamsen et al., 2021).

Circuit-switched and packet-switched technologies can be used to send data over the networks (Suhaimy et al., 2022). Power line communication (PLC) is one alternative for sending messages and data; Digital subscriber lines (DSL) are another communication alternative that is often used in smart metering applications. Wired alternatives include copper wires and fiber optics. There are also many wireless alternatives that can be used for communication (Suhaimy et al., 2022).

Modern smart grid communication allows good communication among the many domains associated with the grid, including marketing, operations, service providers, generation, transmission, distribution, and customers. Home automation is an important feature of the smart grid, which can include charging an EV at low cost rates at night. Home energy management systems may be used to monitor, control, and manage power consumption based on needs and price information (Abrahamsen et al., 2021). The smart grid needs to have clearly defined standards that enable all good communication to reach desired destinations.

The smart grid includes information technology and operation technology. Data such as customer use of electricity and rate when used is information that is important for billing, while electric power production by a solar PV system is operational data that is important for balancing supply and demand. Both the communication systems and the operation of the equipment that provides and distributes electrical power are important to the total system. Communication is included in both systems (Suhaimy et al., 2022). Electrical power flows through the distribution grid to customers, and information flows through the communication network.

Cyber security is a very important issue for the smart grid because the grid needs to be secure in every aspect. Since there are many individuals who are part of the network, there are many communications and large quantities of data that need to be reliable. Billings to customers should be correct, and it is desirable for individuals to be able to verify that their bill is correct. Deliberate attacks on the grid should be able to be detected.

4.5 Benefits of a Smart Grid

There are many benefits of a smart grid (Kabeyi and Olanrewaju, 2023), and this is an effort to provide a summary of them:

1. More opportunities are available to integrate distributed generation into the grid.
2. Good communication is available for demand management implementation.
3. Solar PV-generated electricity can be integrated into the grid efficiently.
4. Time-of-use rates and real-time rates can be implemented effectively.
5. Good communication is available to reduce peak demand when actions are needed.
6. Energy storage can be integrated into the grid to improve stability and efficiency.

7. Hydrogen production from electricity can be integrated into the grid more effectively.
8. A smart grid is essential for meeting the goals of the Paris Agreement to reach net zero greenhouse gas emissions by 2050.
9. There are more effective opportunities for consumers and grid managers to work cooperatively to integrate renewable electricity into the grid.
10. Electric vehicle charging can be accomplished efficiently at a lower cost.
11. Grid operations and maintenance are more efficient, and control is better.

4.6 Dispatchable Sources

One of the benefits of the smart grid is the ability to manage distributed generation, such as solar panels on buildings and in parking lots. Battery storage can be dispatched to provide electricity to the grid when the supply needs to be increased. The dispatch of energy in batteries can include those in EVs and those used at battery swap facilities. Daily dispatch needs can be provided using batteries where there is sufficient renewable energy to charge all of the batteries that are needed.

Hydrogen produced by electrolysis of water using renewable energy can be dispatched to produce electricity using fuel cells. Green hydrogen from renewable energy can be converted to ammonia, which is less expensive to store for later use. Methane produced via anaerobic digestion can also be dispatched to generate electricity.

One of the important alternatives is the shifting of demand to the times of the day when electricity generation from wind and solar is greater (Kabeyi and Olanrewaju, 2023). Demand-side management can be implemented more effectively because of a smart grid that improves communication and choices for customers (Dahiru et al., 2023).

There are dispatchable alternatives, such as standby electricity generating equipment that can be powered by natural gas or mixtures of natural gas and ammonia.

4.7 Standards and Regulations

In order to have reliable and effective operations, many standards and regulations have been adopted and followed (Kabeyi and Olanrewaju, 2023). The importance of interoperability is one of the reasons for adopting and following standards. The National Institute of Standards and Technology (NIST) is one of the more than 25 important standards development organizations.

Since many electric utilities operate as regulated companies, the rate structure that they use is approved by a commission, and the companies have required obligations to follow. The process of developing good regulations and policies is impacted by the complexity of the process.

Public education is needed to educate all who are involved in developing the agreements that are adopted and followed.

References

Abrahamsen, F.E., Ai, Y. and Cheffena, M. 2021. Communication technologies for smart grid: A comprehensive survey. Sensors 21: 8087.

Dahiru, A.T., Daud, D., Tan, C.W., Jagun, Z.T., Samsudin, S. and Dobi, A.M. 2023. A. comprehensive review of demand side management in distributed grids based on real estate perspectives. Environmental Science and Pollution Research 30: 81984–82013.

Dorji, S., Stonier, A.A., Peter, G., Kuppusamy, R. and Teekaraman, Y. 2023. An extensive critique on smart grid technologies: Recent advancements, key challenges, and future directions. Technologies 11: 81. https://doi.org/10.3390/technologies11030081.

Erickson, L.E. and Brase, G. 2020. Reducing Greenhouse Gas Emissions and Improving Air Quality: Two Interrelated Global Challenges. CRC Press, Boca Raton, FL.

Erickson, L.E. and Ma, S. 2021. Solar-powered charging networks for electric vehicles. Energies 14: 966.

Erickson, L.E., Robinson, J., Brase, G. and Cutsor, J., Eds. 2017. Solar Powered Charging Infrastructure for Electric Vehicles: A Sustainable Development. CRC Press, Boca Raton, FL.

Escobar, J.J.M., Matamoros, O.M., Padilla, R.T., Reyes, I.L. and Espinosa, H.Q. 2021. A comprehensive review on smart grids: Challenges and opportunities. Sensors 21: 6978.

Feng, Y. and Lu, X. 2021. Construction planning and operation of battery swapping stations for electric vehicles: A literature review. Energies 14: 8202.

Hasankhani, A., Hakimi, S.M., Bisheh-Niasar, M., Shafie-Khah, M. and Asadolahi, H. 2021. Blockchain technology in the future smart grids: A comprehensive review and frameworks. International Journal of Electric Power & Energy Systems 129: 106811.

Kabeyi, M.J.B. and Olanrewaju, O.A. 2023. Smart grid technologies and application in the sustainable energy transition: A review. International Journal of Sustainable Energy 42: 685–758.

Korcer, M.C., Onen, A., Ustun, T.S. and Albayrak, S. 2022. Optimization of multiple battery swapping stations with mobile support for ancillary services. Frontiers in Energy Services. https://doi.org/10.3389/fenrg.2022.945453.

Suhaimy, N., Radzi, N.A.M., Ahmad, W.S.H.M.W., Azmi, K.H.M. and Hannan, M.A. 2022. Current and future communication solutions for smart grids: A review. IEEE Access 10: 1109.

Sultan, V., Aryal, A., Chang, H. and Kral, J. 2022. Integration of EVs into the smart grid: A systematic literature review. Energy Informatics 5: 65.
Wu, H.2022. A survey of battery swapping stations for electric vehicles: Operation modes and decision scenarios. IEEE Transactions on Intelligent Transportation Systems 23: 10163–10185.

5

Reliability

5.1 Introduction

Our present electrical grid system is generally very reliable (energy is consistently available when wanted) and resilient (energy is available even when the system is under stress or otherwise challenged). Reliability is something that people are used to, that people (and companies) rely on, and that needs to be maintained. Reliability and resilience are, therefore, very important in the pathways toward net zero greenhouse gas emissions. This chapter focuses on reliability, which tends to imply resilience in that resilience is often the ability of a system to continue to perform (i.e., be reliable) when under stress.

Our present power grid system is very good, but not perfect. For example, in February of 2021, a severe winter storm impacted Texas and several other states, and the Texas system revealed itself to be not entirely reliable or very resilient. Many generating units ceased operating because of the storm. In Texas, approximately 34,000 MW of generating capacity was unavailable for about two days, which impacted around 4.5 million people. As a result of this electrical infrastructure failure, over 200 people died, and there was about $100 billion in direct and indirect losses in Texas (FERC, 2021).

Across multiple surveys of people (by one of the authors, GB) about what their attitudes, concerns, and likely decisions are about different types of energy sources (solar, fossil fuels, etc.), a top concern that repeatedly shows up is keeping energy costs low. In discussing these findings with executives in energy distribution companies, they also agreed that one major goal was to keep energy prices as low as possible—but they also pointed out that reliable, robust energy was another major goal. From the distributor's perspective, it was a fantastic achievement, although perhaps a bit frustrating, that customers generally seem to assume their energy will reliably be there even under challenging conditions.

There were actually two reliability aspects associated with the 2021 Texas energy grid collapse. One aspect was the reliability of the electrical grid in providing electricity. The other was the reliability of each customer to have what is needed to address the electrical power outage without loss of life and significant property damage because of frozen pipes and other impacts. If the first aspect of reliability (grid reliability) is high, then the second aspect of reliability

DOI: 10.1201/9781003396154-5

(functional point-of-use reliability) is not as important. If grid reliability is not great, then the latter aspect—point-of-use reliability—becomes very important. The 2021 Texas storm is again a good example, but there are many events each year where electric power is off. The number of significant power failure events each year has been increasing because of climate change.

Reliability must be included in the development of pathways to net zero greenhouse gas emissions. The issue of reliability includes, as part of its domain, the issue of resilience. It is also possible, though, to carve up the issue of reliability in terms of the grid reliability in providing electricity and the ability for consumers to realize point-of-use reliability.

5.2 Electric Grid Reliability

The reliability of electrical grids has been a perennial concern in most locations. In the United States, there are ongoing efforts to address reliability issues and improve reliability that are actually formalized. The mission of the North American Electric Reliability Corporation (NERC) and the six associated regional organizations is to "assure the effective and efficient reduction of risks to the reliability and security of the grid" (NERC, 2023). Professionals from NERC contributed to the recent report on the disaster in Texas (FERC, 2021). Since 2021, NERC has published a 2022 long-term reliability assessment that addresses risks associated with severe winter storms, such as the storm that impacted Texas in February 2021 (NERC, 2022). This report provides recommendations for utilities in order to prepare for extreme weather demands. With the transition to more generation of electricity from wind and solar, there is a need to plan for the integration of renewable power into the risk analysis. Wind and solar are not dispatchable and may not be available in a cold winter period because of snow on the solar panels or ice on the wind turbine blades. Under normal weather conditions, wind and solar may contribute much of the electricity that needs to be generated at prices that are less than the cost of fuel for coal plants and natural gas plants. Severe weather conditions, though, need to be considered.

There may be a need to keep old coal plants operational for severe weather conditions, which includes the need to be able to prevent problems such as a frozen coal pile (although nuclear or natural gas plants would be better, environmentally). With the transition to electric vehicles and buildings with zero greenhouse gas emissions, more electricity needs to be generated daily, and the need to manage winter peak power becomes more important. In some respects, there is, therefore, greater uncertainty because of the transition to more electricity being used for heating as well as the uncertainty associated with the amount of electricity from wind and solar generation.

5.3 Point-of-Use Reliability

Short of a perfectly reliable power grid, it is possible to make functional reliability better at the location where power is being used. Many people are familiar with this in the context of places like hospitals, where it is essential that there is reliable electricity even if the power grid fails. A traditional point-of-use reliability option that has been typically implemented is the installation of a generator that provides electricity when there is a power failure. Historically, fossil fuels (gas or diesel) have been used to power a motor that provides temporary power. A residential house that needs to have good alternatives to power grid reliability (e.g., to have power during an electrical outage) can also use a fossil fuel-powered generator.

Moving forward, renewable energy sources like solar and wind power, especially in coordination with battery storage, can make point-of-use reliability much more common and achievable. Battery storage is a growing option for both reducing risk in the management of power and in building reliability of the power grid. With increases in the amount of electrical power generated by wind and solar and the increases in battery capacity, the potential to use batteries for reliability purposes is increasing. Battery storage in buildings can be used for emergency electrical needs and can provide point-of-use reliability. These backup batteries can be purpose-built and installed, but there is also work on how to repurpose the large batteries from some EVs as home backup power sources. An EV battery can have sufficient energy to provide emergency service to a home for several days. In fact, it is now possible to connect some EVs to home power systems and use the vehicle battery as a power source for the home as a backup.

Of course, combining point-of-use renewable power generation and batteries can yield multiple benefits above and beyond either approach by itself. An on-site generator powered by, say, solar energy can generate power day after day without needing to be refilled with a fossil fuel. Connecting that power generator to a battery provides more reliability—energy both when the generator is working and (from the battery) when it is not. And, of course, this system would be part of the solution in the transition to net zero greenhouse gas emissions.

Solar power is not the only option for point-of-use power generation and reliability. There are many options for fuels that have no greenhouse gas emissions, such as green hydrogen and ammonia from renewable electricity and others from renewable sources, such as methane from anaerobic digestion. Biomass can be used as a green fuel or processed to obtain a fluid fuel.

Hydrogen can be used to produce electricity using a fuel cell, which can be more efficient than using hydrogen as a green fuel.

References

FERC. 2021. The February 2021 Cold Weather Outages in Texas and the South Central United States. Federal Energy Regulatory Commission Report, November 2021. ferc.gov/

NERC. 2022. 2022 Long-Term Reliability Assessment. North American Electric Reliability Corporation, Washington, DC. nerc.com/

NERC. 2023. 2023–2024 Winter Reliability Assessment. North American Electric Reliability Corporation, Washington, DC. nerc.com/

6

The Electrification of Transportation

6.1 Introduction

There has been great progress in the electrification of transportation since 2000 when the first commercial hybrids, the Toyota Prius and Honda Insight, became widely available. In 2024, there are many good choices for people to choose from, including electric sedans, hatchbacks, sportscars, and pickup trucks. There are also electric bicycles, scooters, motorcycles, buses, and commercial trucks. Enough choices exist on the market now for good competition and even some price reductions to encourage sales.

The total number of electric vehicles (EVs) in service is expected to grow to about 250 million in 2030 and then 585 million in 2035, which is a growth rate of about 23% annually (IEA, 2024). Those numbers exclude two- and three-wheel EVs, which will be covered separately. China is the leader in the number of EVs in service in 2024 and may reach 50% of new car sales being EVs before 2030 (IEA, 2024). In 2030, the top 20 automobile manufacturers are expected to be up to 42% EVs or more in their combined new car sales (IEA, 2024).

The move to EVs has not been entirely smooth, though. Imbalances of supply and demand curves led first to a shortage of EVs relative to consumer demand and then oversupplies of EVs more recently. In 2023 and 2024, there were price reductions of 5–10% on some EVs to improve sales (IEA, 2024). BYD and Tesla are currently the leaders in market share for 2023 and 2024, with 35% of EV sales. BYD was the global market leader in new car EV sales in 2023 while Tesla had the most EV sales in the U.S. (IEA, 2024).

It is well known that EVs are much more efficient than internal combustion engines and that electric motors do not have emissions (Erickson and Brase, 2020; Muratori et al., 2021). Additionally, EVs have just a fraction of the moving parts found in an internal combustion engine vehicle. That means less maintenance over the life of the vehicle: no oil changes, no transmission issues, no fuel system maintenance, and no exhaust system issues. All this means that, even with EVs that cost more for an initial purchase, the total cost of ownership can be less than with traditional vehicles. With low-cost renewable electricity from wind and solar sources, operating costs are even less.

DOI: 10.1201/9781003396154-6

There are many good sources that describe progress in the electrification of transportation (Erickson et al., 2017; Erickson and Brase, 2020; Erickson, 2024). This chapter will emphasize recent developments and the state of progress in 2024. As we will see, China, the United States, Europe, and many other parts of the world have made significant progress in the electrification of transportation.

6.2 Types of Electric Cars

In 2000, the Toyota Prius operated with an electric battery motor that was charged using power from a gasoline engine and the momentum of the vehicle. This type of EV is a hybrid electric vehicle (HEV). The efficiency of HEVs is better than that of the same model with only a gasoline engine. The plug-in hybrid Toyota Prius (PHEV) has a larger battery, which provides the energy needed to have an all-electric range, but still retains a gasoline engine for travel beyond that range. The Nissan Leaf is an example of a battery electric vehicle (BEV) that has a battery and electric motor with no gasoline engine at all. BEVs, unlike HEVs, cannot continue after the battery charge is exhausted, and the current trend is for BEVs with increasingly longer ranges (now typically about 250–300 miles). One of the other EV options that may come into play in the future is the fuel cell EV, which uses hydrogen in a fuel cell to generate electricity that powers the vehicle.

In 2023, global sales of electric cars increased to 18% of all new car sales, constituting more than 13 million EV sales. More than 50% of these EVs were produced in China (IEA, 2024). About 60% of EV sales in 2023 were also in China, compared to 25% in Europe and 10% in the USA. These three geographical regions had about 95% of EV sales and about 65% of all car sales because there has been the greatest progress in the electrification of transportation in these three regions.

In 2022 and 2023, there was notable progress in reducing the prices of EVs in many countries, such that new car prices for EVs are lower than prices for conventional cars with internal combustion engines (ICEs). The total cost of ownership (including purchase, fuel, and maintenance) is less for EVs in many countries in 2023. Additionally, China and India have made progress in developing and marketing inexpensive EVs (Erickson, 2024; IEA, 2024). This involves building simple, basic EVs with modest ranges that work very well in urban settings.

Many people have found that BEV ownership is their preference for local driving and that it is economical and convenient. They can charge the battery at home, and the operating costs are low. For people who want to help reduce greenhouse gas emissions, this is one way to do so.

Those who wish to take a step further can generate their own electricity by installing solar panels at home and charging their EV with their own renewable electricity.

Those who own a PHEV can drive locally on just electricity. One author (LEE) has owned a plug-in hybrid Toyota Prius since August 2013, which now has over 99,000 miles of service as of April 2024. The 12 miles of electric range on this Prius, with the 4.4 kWh battery, is sufficient for daily driving locally on most days. A PHEV can also, of course, go very long distances by using the gasoline motor to charge the EV battery. When driving in the mountains, the author has actually been able to watch the battery getting charged simply from the momentum while driving downhill.

A great deal of focus over the past few years has been on EV cars and trucks as personal vehicles. There is much more going on, however, with regard to the overall electrification of transportation. The following sections review progress on electric buses and commercial trucks, bicycles, scooters, and motorcycles. Overall, there has been good progress in the development of many new products for the electrification of transportation (IEA, 2024).

6.2.1 Electric Buses and Trucks

There are eight classes of vehicles based on weight, according to the U.S. Federal Highway Administration (Fleming et al., 2021):

Class 1—up to 6,000 lbs;
Class 2—6,001 to 10,000 lbs;
Class 3—10,001 to 14,000 lbs;
Class 4—14,001 to 16,000 lbs;
Class 5—16,001 to 19,500 lbs;
Class 6—19,501 to 26,000 lbs;
Class 7—26,001 to 33,000 lbs;
Class 8—33,001 lbs and greater.

Sales of Class 2 and heavier electric trucks are much smaller in number compared to Class 1 sales, which include personal use electric cars (IEA, 2024).

There has been good progress in the electrification of commercial trucks (IEA, 2024). In 2023, light commercial vehicle sales (Class 2+) were greater than 240,000 in China and about 148,000 in Europe, while total global electric truck sales were about 54,000 for large trucks (IEA, 2024). Most of the sales were in China, Europe, and North America. There are more than 100 original equipment manufacturers developing electric trucks and buses globally (IEA, 2024). The number of sales increased in 2023 compared to 2022. Global electric truck sales are projected to be 30% of sales in 2035 (IEA, 2024). At COP 28, the number of countries that have pledged to participate in the plan to have sales of only zero-emission electric trucks after 2039 was up to 33 (IEA, 2024).

There are about 635,000 electric buses in service globally as of December 2023. New electric bus sales were about 50,000 in 2023. In China, about 25% of buses that are in service are electric as of 2024. City buses are the most popular market because of their ability to manage their operations and charging process. In the USA, there are both public transportation buses in cities and school buses powered by electricity. Many European countries have electric buses, and China exports electric buses to many countries (IEA, 2024). The estimate is that 30% of buses sold globally will be EVs by 2035 (IEA, 2024). Many cities and countries are expected to add electric buses between 2024 and 2035, including Buenos Aires, Taipei, Chile, Colombia, Ecuador, the Philippines, the Dominican Republic, Nepal, Pakistan, and Panama (IEA, 2024).

As part of an initiative to accelerate the adoption of EVs, the US federal government has been offering grants to school districts to replace their buses with electric versions. Nearly 400 school districts across all 50 states and Washington D.C. applied for this funding. According to the EPA, about 2,000 applications were received, requesting nearly $4 billion for more than 12,000 buses. Because funding was limited, 389 applications (totaling $913 million) were approved for the purchase of 2,463 buses (Daly, 2022)

Given the smaller numbers of vehicles in these categories, why pay so much attention to them? There are a couple of reasons to focus on “working” vehicles in general. One reason is that they are in much more continuous use compared to a private vehicle, which is often parked for about 22 hours every day. A second reason is the potential visibility of EVs, given the amount of time they are being used and that they are commonly in high-traffic areas. A third reason is that, based on the nature of how these vehicles are used, EVs can be particularly useful. Many service vehicles, such as city buses, school buses, postal delivery vans, and garbage trucks, run on fixed routes. A vehicle on a fixed route can be fitted with a battery that provides the range needed to cover that route. Other service vehicles operate based more on an amount of time, such as a work shift, and the needed range is less precise but can still be estimated fairly well. This includes package delivery vans (e.g., FedEx, UPS, and DHL), taxis, police cars, ambulances, and fire trucks. Most fleet systems are good candidates for transitioning to EVs, which include government vehicles ranging from meter readers to road maintenance vehicles to animal control vehicles to safety inspector vehicles.

Finally, as noted in a New York Times article (Heilweil, 2022), replacing working vehicles that are in more or less continuous use can have an outsized impact on reducing carbon emissions. Although fleet delivery vehicles represented less than 10% of registered vehicles in 2020, they accounted for 26% of the transportation sector’s greenhouse gas emissions.

With EV adoption in situations where the vehicles are in very high use, there can be a good argument for battery swapping systems. This allows a vehicle to be used for a shift with one battery, then quickly used again with a new battery (and, perhaps, a new human on the next shift). There has been progress in the developments in battery swapping technology for charging electric trucks and buses in 2023. Many heavy-duty trucks and buses are able to make use of battery-swapping technologies (IEA, 2024).

6.2.2 Electric Bicycles, Motorcycles, and Scooters

The number of electric bicycles, motorcycles, and scooters is burgeoning in China, India, and across the Association of Southeast Asian Nations (ASEAN). These are good markets for both two-wheel and three-wheel EVs. In 2023, 30% of two- and three-wheel sales were EVs in China compared to 8% in India. In 2023, sales were almost 6 million two-wheel EVs in China, 880,000 in India, followed by 380,000 in ASEAN countries (IEA, 2024). In India, the growth in two-wheel EV sales was 40% compared to 2022.

Nearly 1 million new electric three-wheel vehicles were delivered and put in service in 2023. The global increase in sales of three-wheel EVs was 30% compared to 2022. India is the largest market because of the popularity of the electric auto-rickshaw, which is the most cost-effective alternative compared to the ICE rickshaw. About 25% of three-wheel vehicles were EVs in 2023 (IEA, 2024). The market in China for three-wheel EVs is also large (IEA, 2024). Just about everywhere, the operational costs are lower for electric three-wheel vehicles compared to the ICE alternative.

The ASEAN countries have a larger proportion of scooters compared to China and India. Vietnam is the leading country in the production of two-wheel EVs among the ASEAN countries and now exports vehicles to other countries. Indonesia is increasing its manufacturing activities to produce more electric bicycles (IEA, 2024). In total, there are now about 65 million two- and three-wheel EVs in service globally as of 2024, and the projected numbers are 210 million in 2030 and 360 million in 2035 (IEA, 2024).

6.2.3 Electric Trains and Planes

The electrification of trains is a topic for some locations, such as the United States, but much of the rest of the world already has trains (and trams) that mostly run on electricity. The easiest model for the U.S., therefore, is simply to move towards the electric systems for trains that have been developed in the rest of the world.

The electrification of airplanes poses a more substantial and worldwide challenge. The amount of force required to take off, maintain altitude, and do so for long periods of time is tremendous. Just adding more batteries is not a good option because airplanes need to weigh less, not more, to be efficient. Currently, the only common battery-powered aircraft are ultra-light, single-passenger, and used just for short distances. Things may change in the future, though. Wolleswinkel and colleagues (2024) have worked out how to build, in principle, battery-electric aircraft that can replace existing larger aircraft. Their approach involves re-assessing some core aspects of aircraft engineering (standard Class-I versus Class-II mass and aerodynamic efficiency estimates), but with these changes, there appears to be a design space where battery-electric passenger aircraft can work. They even have provided the design details for a 90-seat aircraft with a battery-powered useful range of 800 km (de Vries et al., 2024).

6.3 Vehicle Charging

One of the key issues for EV owners and operators is the need to have electricity to charge their EV. Actually, this may be more of a key issue for prospective EV owners than for those who already have an EV. Pretty much all EVs can be charged at home using a level 1 system, which is a standard 120-volt wall outlet. This provides about 20 amps of current and 2.4 kW of power in the USA. For many EVs, this works well when charging after local daily driving. One way to think about it is that your car (as an EV) becomes just like your phone; when you are done with it for the day, you plug it in to charge overnight. In the morning you have a full charge again.

One of the authors (GLB) has owned a BEV Nissan Leaf since 2015, which has about 75 miles of range. With home charging (Level 1, 120 volts), that range has been sufficient for daily driving around town for the past 10 years, with the exception of five occasions when he has used a public charging station. For long trips (out of town), of course, he uses his spouse's ICE vehicle, but that will not be necessary if he ever buys any of the newer BEVs with a much longer range.

Some owners have a dedicated system for charging their electric vehicle, which usually involves setting up a specific plug in the garage with Level 2 (240 volt) charging. (Nearly all homes have 240-volt availability for major appliances such as air conditioning units and washers/dryers.) Level 2 charges more quickly, which can be appealing if the EV is being driven for substantial distances (e.g., over 100 mile/day) on a regular basis. This can maintain the ability to completely recharge overnight, and with this amount of driving, there are probably enough fuel/maintenance savings to justify the dedicated charging system.

An even faster way to get to a full battery is to swap a depleted battery for a full one: battery swapping. There has been significant progress in the development of battery swapping technology because of the advantages of reduced time to accomplish the process and the ability to charge batteries at a slower rate and at a reduced cost (Arora et al., 2023; Cui et al., 2023; Qiang et al., 2023; Revankar and Kalkhambkar, 2021). Battery swapping is now being used at many locations in China.

With the number of EVs on the road increasing every year, there are continued developments in home charging, public battery charging services, and battery swapping. All of these need to operate within the context of an electric grid, and as we discussed in Chapters 4 and 5, that electrical grid has to be able to manage the supply and demand fluctuations entailed by more power generation from solar, wind, and other renewable sources. Because of progress in the generation of electricity using wind and solar sources, there will sometimes be a need for on-demand supplemental power sources, and there will also sometimes be excess energy generated, which needs to be either used immediately or stored. In short, there is a greater need for smart grids with electricity demand management.

From one perspective, the addition of more EVs simply means more demand is placed on the electrical grid. The existence of more EVs, however, can also help with grid demand management. Consider that many private car EVs charge overnight, which is when other electrical demands are usually low; this evens out the overall demand curve. If EV owners are flexible about when their vehicle charges (and perhaps want to get a discount price), a smart electrical grid can decide when to provide EV charging based on other supply/demand considerations. If owners are able to plug their EV directly into an energy source, such as solar panels installed over their parking lot, the grid is essentially bypassed. An advantage of battery swapping systems is the greater opportunity to charge batteries when the electricity supply/demand proportion is highest.

6.4 Demand Management

There is a need for effective demand management to always make good use of all of the electricity that is generated from renewable sources. The generation of electricity using wind and solar resources is on a pathway of tremendous expansion, both to take advantage of lower production costs and to reduce greenhouse gas emissions. This is part of the plan to reach net zero by 2050.

One aspect of demand management is to charge batteries when supply from renewable energy sources is particularly high, and demand thus needs to be increased. Some of this additional demand can be batteries in EVs. For instance, solar panels in parking lots can be connected to both parked EVs and the grid; batteries in EVs and batteries at swap stations can be charged at reduced prices as part of a demand management program to help increase demand to balance supply and demand in grids where there are many renewable sources. This is a dynamic process that can be managed by the grid operator, automated systems, and others helping with demand management (Erickson and Ma, 2021).

With greater electrification of the transportation sector comes more demand for electricity. This greater demand, if managed well, can be met by expanded energy generation from renewable sources. With expanded renewable energy production, there are more opportunities for further electrification in the transportation (and other) sectors. Properly designed, a virtuous cycle can be created that helps overall. This is similar to the positive interaction effects of combining EVs and solar energy generation (Erickson et al., 2017; Erickson and Ma, 2021)

References

Arora, A., Murarka, M., Rakshit, D. and Mishra, S. 2023. Multiobjective optimal operation strategy for electric vehicle battery swapping station considering battery degradation. Cleaner Energy Systems 4: 100048.

Cui, D., Wang, Z., Liu, P., et al. 2023. Operation optimization approaches of electric vehicle battery swapping and charging station: A literature review. Energy 263: 126095.

Daly, M. 2022. More kids to ride in "clean" school buses, mostly electric. Oct 27. https://apnews.com/article/business-kamala-harris-seattle-washington-pollution-16405c66d405103374d6f78db6ed2a04

de Vries, R., Wolleswinkel, R.E., Hoogreef, M. and Vos, R. 2024. A New Perspective on Battery-Electric Aviation, Part II: Conceptual Design of a 90-Seater. In Proceedings of the AIAA SCITECH 2024 Forum [AIAA 2024–1489] American Institute of Aeronautics and Astronautics Inc. (AIAA).

Erickson, L.E. 2024. Electric vehicles for environmental sustainability. In Decarbonization Strategies and Drivers to Achieve Carbon Neutrality for Sustainability, M.N.V. Prasad, L.E. Erickson, F.C. Nunes, and B.S. Ramadan, Eds. Elsevier, Amsterdam.

Erickson, L.E. and Brase, G. 2020. Reducing Greenhouse Gas Emissions and Improving Air Quality: Two Interrelated Global Challenges. CRC Press, Boca Raton, FL.

Erickson, L.E. and Ma, S. 2021. Solar powered charging networks for electric vehicles. Energies 14: 966.

Erickson, L.E., Robinson, J., Brase, G. and Cutsor, J. 2017. Solar Powered Charging Infrastructure for Electric Vehicles: A Sustainable Development. CRC Press, Boca Raton, FL.

Fleming, K.L., Brown, A.L., Fulton, L. and Miller, M. 2021. Electrification of medium- and heavy-duty ground transportation: Status report. Current Sustainable/Renewable Energy Reports 8: 180–188.

Heilweil, R. 2022. The Future of EVs Isn't Your Tesla: It's Rental Cars and Buses. New York Times.

IEA. 2024. Global EV Outlook 2024. International Energy Agency. www.iea.org

Muratori, M., Alexander, M., Arent, D., et al. 2021. The rise of electric vehicles: 2020 status and future expectations. Progress in Energy 3: 022002.

Qiang, H., Hu, Y., Tang, W. and Zhang, X. 2023. Research on optimization strategy of battery swapping for electric taxis. Energies 16: 2296.

Revankar, S.R. and Kalkhambkar, V.N. 2021. Grid integration of battery swapping station: A review. Journal of Energy Storage 41: 102937.

Wolleswinkel, R.E., de Vries, R., Hoogreef, M.F.M. and Vos, R. 2024. A New Perspective on Battery- Electric Aviation, Part I: Reassessment of Achievable Range. In Proceedings of the AIAA SCITECH 2024 Forum [AIAA 2024–1489] American Institute of Aeronautics and Astronautics Inc. (AIAA). https://doi.org/10.2514/6.2024-1489.

7

Heating, Cooling, and Lighting Buildings

7.1 Introduction

One of the important sources of greenhouse gas emissions is the emissions from buildings associated with heating, cooling, lighting, and other activities that generate emissions in buildings. About 36 to 40% of greenhouse gas emissions come from buildings (Papadakis and Katsaprakakis, 2023; Hashempour et al., 2020). Many homes are heated with natural gas or fuel oil, and many water heaters burn natural gas.

7.2 Heating and Heat Pumps

Historically, heating has been provided by burning natural gas in furnaces in many locations because of the low cost of the fuel. The cost of energy from the electrical grid comparatively has been much greater than the cost of natural gas. The difference in cost is an important issue. One way to reduce the use of energy for heating and cooling in buildings is to improve insulation. This is a relatively fast, easy, and inexpensive way to be more efficient and save money. Although it does not end the use of fossil fuels for heating, it uses less fuel.

One pathway to net zero in the realm of heating is to install heat pumps for heating because of their higher efficiency. Heat pumps can be powered by renewable electricity to reduce greenhouse gas emissions. They can also be used for both heating and cooling. Heat pumps have a significant history related to their development and use because of the better energy efficiency that is associated with their operation. The coefficient of performance (COP) is the ratio of the heat provided divided by the work required to power the heat pump. Values of COP vary with ambient outdoor temperature; the values are often between 1.5 and 3.5, with smaller values for colder ambient temperatures.

There are a number of alternatives to consider when selecting heat pumps. Air source heat pumps use the outside air in the operation of the system (Carroll et al., 2020). Ground source heat pumps are a good alternative in some locations because the ground temperature is warmer compared to the outside

DOI: 10.1201/9781003396154-7

air temperature (Lu and Ziviani, 2022). In cold regions, there are alternatives, such as using two-stage compression systems (Zhang et al., 2018). Heat pumps can also be used to heat water for warm water needs.

The current cost of heating buildings depends on local costs of natural gas, fuel oil, and electricity, as well as the temperature of the outside air. Eight different configurations have been compared at 11 different U.S. locations using EnergyPlus computer simulation (Lu and Ziviani, 2022). The local costs for natural gas and electricity for each location, as well as local weather data, were used. In some locations, natural gas heating using a furnace and gas water heater gave the lowest cost; however, there were some locations where the heat pump alternative provided the lowest cost.

The progress in reducing the cost of renewable electricity generation using wind and solar processes is important with respect to the transition to all-electric buildings. Time-of-use prices for electricity and demand management alternatives for building operations can be used to improve the economics of heat pumps in buildings.

One of the concerns in the transition to heat pumps is major winter snow storms because they create a need for heat in buildings and reduce the effectiveness of generating electricity using wind and solar energy. The electrical grid needs to have a mix of alternatives to provide electricity under these conditions.

7.3 Solar Heating and Hot Water

Another pathway to lower emissions from buildings is the use of solar energy for heating buildings and/or water. At the present time, natural gas is used to heat buildings and provide hot water in many locations.

There are many installations of solar hot water heaters, and there are good publications on this topic (Ahmed et al., 2021; Rahimi-Ahar et al., 2023). Hot water is important in residential buildings, office buildings, stores, and industrial operations. Solar energy is free; however, solar water heaters that use solar energy to heat water must compete with other alternatives. One important issue is the freezing of water in winter. Locations where the temperature is always above zero Celsius do not need to consider this issue. Water can be used as the fluid that is heated; the alternative is to use a fluid that does not freeze and a heat exchanger to heat the water.

The geographical conditions, solar radiation, and temperature are important; thus, solar hot water heaters are better choices in locations where temperatures are higher and solar radiation values are good on most days. Turkey, some parts of Africa, India, and Taiwan are examples of good locations for solar hot water heaters. In the U.S., there are many locations where solar hot water heaters can be used. The southwest has very good solar radiation and good temperatures.

Many issues are considered by those who evaluate the option of using solar energy to heat water. Adoption decisions have been correlated with economic and technical variables that are often considered by those who consider the installation of a solar water heater (Zare et al., 2021). Geographic location was an important variable. Features of the residence were important also. More than 100 variables were identified.

Solar process heat has many applications in industry. The food industry has many washing and cleaning operations where hot water is used. The term Solar Heat for Industrial Processes (SHIP) is used. As of the end of 2020, there were about 891 SHIP projects globally where solar heating is being used. More than one million square meters of solar collection area are installed (Tasmin et al., 2022). A significant portion of the SHIP projects are food and/or beverage plants. Process heating, pasteurization, cooking, and cleaning are important operations where solar energy is used. In U.S. manufacturing, 70% of process energy is used for heating (Schoeneberger et al., 2020). Solar thermal heating, solar PV-generated electricity used for resistance heating, and heat pump heating are all found in industrial applications. The transition to greater use of solar energy in industry is in progress, and new solar thermal heating installations can contribute to the efforts to reach net zero.

Greenhouses make use of solar energy to provide heat and light to manage the desired conditions for horticultural operations. Providing controlled conditions is important for food production in many countries. Temperature, relative humidity, light, vapor pressure deficit, and carbon dioxide concentration are important variables in greenhouses. Crop yields in controlled environments have the potential to improve food production and the reliability of food supply. Renewable energy to operate the system can help reduce greenhouse gas emissions. There is a very good review of controlled-environment agriculture (Vatistas et al., 2022). A greenhouse can be installed on a flat roof in an urban area to produce vegetables and make use of urban land to grow crops.

7.4 Cooling with Renewable Energy

Renewable energy can be used to provide electricity for cooling, and solar panels are very effective at generating electricity on hot summer days. In terms of reducing issues such as building new power plants, transmission lines, and other infrastructure needs, using solar PV-generated electricity from the roof of the building to cool that building is a logistically simple system. Another option, although less well known, perhaps in part because of its name, is that heat pumps can actually be used for both heating and cooling.

Yet another option for thermal comfort is to circulate air in a building. A home with a basement, for example, may have a lower temperature at that lower level. There may be many times when simply circulating the air across floors can improve the temperature on an upper level, allowing the air conditioner to remain shut off.

7.5 Lighting

Advances in lighting are being made in both more efficient artificial lighting and using natural lighting more effectively. Often lighting can be improved by selecting LEDs that are very efficient and last longer. Occupancy sensors that turn lights out when rooms are not occupied will also save energy. More broadly, educating people about how to save energy and the amount of savings that they can achieve is beneficial. Even small changes, such as reminder messages to turn the lights off when leaving a room, can be beneficial.

Solar lighting (daylighting or natural lighting) in buildings has significant value, including health benefits (Osibona et al., 2021), the well-being of occupants (Morales-Bravo and Naverrete-Hernandez, 2022), and work performance (Vasquez et al., 2022). Windows provide both views of the outside and natural light; there are positive benefits of having windows in homes, schools, and offices. Ultraviolet solar radiation in natural light damages bacteria and viruses, and this has health benefits (Hockberger, 2000; Osibona et al., 2021; Sharun et al., 2021). There is significant literature on the benefits of natural light (Mardaljevic, 2021). Vitamin D provided by solar radiation is also a health benefit.

7.6 Conclusions

Buildings are associated with a large proportion of greenhouse gas emissions, but great strides can be made in both increasing the efficiency of structures and shifting to new technologies. Many buildings were built decades ago, and the insulation needs to be improved, together with the installation of better doors and windows to improve efficiency (Erba and Barbieri, 2022). The efficiency of buildings can be improved by making technological improvements, operational changes, and behavioral invitations (Papadakis and Katsaprakakis, 2023). The efficiency of the equipment used for heating and air conditioning can be improved. Energy management systems that change thermostat settings with time of year and with room use can improve efficiency.

Significant progress has been reported on building integrated solar energy systems that include thermal, photovoltaic, and hybrid installations consisting of both (Bot et al., 2022). Near zero energy buildings that make use of sunlight and renewable energy with the integration of electricity from renewable sources and thermal energy from solar sources are included in this review. A number of different building-integrated solar energy systems are described.

References

Ahmed, S.F., Khalid, M., Vaka, M., et al. 2021. Recent progress in solar water heaters and solar collectors: A comprehensive review. Thermal Science and Engineering Progress 25: 100981.

Bot, K., Aelenei, L., Gomes, M.D.G. and Silva, C.S. 2022. A literature review on Building Integrated Solar Energy Systems (BI-SES) for facades- photovoltaic, thermal and hybrid systems. Renewable Energy and Environmental Sustainability 7: 2021053.

Carroll, P., Chesser, M. and Lyons, P. 2020. Air source heat pump field studies: A systematic literature review. Renewable and Sustainable Energy Reviews 134: 110275.

Erba, S. and Barbieri, A. 2022. Retrofitting buildings into thermal batteries for demand-side flexibility and thermal safety during power outages in winter. Energies 15: 4405.

Hashempour, N., Taherkhani, R. and Mahdikhani, M. 2020. Energy performance optimization of existing buildings: A literature review. Sustainable Cities and Society 54: 101967.

Hockberger, P.E. 2000. The discovery of the damaging effect of sunlight on bacteria. Journal of Photochemistry and Photobiology B: Biology 58: 185–191.

Lu, Z. and Ziviani, D. 2022. Operating cost comparison of state-of-the -art heat pumps in residential buildings across the United States. Energy and Buildings 277: 112553.

Mardaljevic, J. 2021. The implementation of natural lighting on human health from a planning perspective. Lighting Research and Technology 53: 489–513.

Morales-Bravo, J. and Navarrete-Hernandez, P. 2022. Enlightening wellbeing in the home: The impact of natural light design on perceived happiness and sadness in residential spaces. Building and Environment 223: 109317.

Osibona, O., Solomon, B.D. and Fecht, D. 2021. Lighting in the home and health: A systematic review. International Journal of Environmental Research and Public Health 18: 609, http://doi.org/10.3390/ijerph18020609.

Papadakis, N. and Katsaprakakis, D.A. 2023. A review of energy efficiency interventions in public buildings. Energies 16: 6329. https://doi.org/10.3390/en16176329.

Rahim-Ahar, Z., Khiadani, M., Ahar, L.R. and Shafieian, A. 2023. Performance evaluation of single strand and hybrid solar water heaters: A comprehensive review. Clean Technologies and Environmental Policy 25: 2157–2184.

Schoeneberger, C.A., McMillan, C.A., Kurup, P., Akar, S., Margolis, R. and Masanet, E. 2020. Solar for industrial process heat: A review of technologies, analysis, approaches, and potential applications in the United States. Energy 206: 118083.

Sharun, K., Tiwari, R. and Dhama, K. 2021. COVID-19 and sunlight: Impact on SARS-CoV-2 transmissibility, morbidity, and mortality. Annals of Medicine and Surgery 66: 102419.

Tasmin, N., Farjana, S.H., Hossain, M.R., Golder, S. and Mahmud, M.A.P. 2022. Integration of solar process heat in industries: A review. Clean Technologies 4: 97–131.

Vasquez, N.G., Rupp, R.F., Andersen, R.K. and Toftum, J. 2022. Occupants' responses to window views, daylighting, and lighting in buildings: A critical review. Building and Environment 219: 109172.

Vatistas, C., Avgoustaki, D.D. and Bartzanas, T. 2022. A systematic literature review on controlled-environment agriculture: How vertical farms and greenhouses can influence the sustainability and footprint of urban microclimate with local food production. Atmosphere 13: 1258.

Zare, S.G., Hafezi, R., Alipour, M., Tabar, R.P. and Stewart, R.A. 2021. Residential solar water heater adoption behavior: A review of economic and technical predictors and their correlation with the adoption decision. Energies 14: 6630.

Zhang, L., Jiang, Y., Dong, J. and Yao, Y. 2018. Advances in vapor compression air source heat pump system in cold regions: A review. Renewable and Sustainable Energy Reviews 81: 353–365.

8

Agriculture and Global Food Systems

8.1 Introduction

Much of the earth is used for agriculture: producing the food that feeds billions of people who live there. Agriculture is, therefore, another important topic that must be addressed in the goal of reaching net zero, and reducing greenhouse gas emissions is needed along the entire food supply system. On farms and ranches, the electrification of equipment and the shift to using renewable nitrogen fertilizer are both very significant issues. Further along the food supply chain, food waste is a significant issue that should be managed better. Farming also involves the intensive use of water resources, so we also need to address the use of water in reservoirs, lakes, rivers, seas, and even oceans.

Although there are many issues in agriculture about greater efficiency and conservation in current practices, there are also tremendous opportunities. The production of renewable electricity and carbon sequestration are possible dual uses of the land used for agriculture. In principle, net zero ethanol can be produced for use as a liquid fuel, and once the transitions have been made to electric tractors and net zero fertilizer, there is a very real pathway to the agricultural sector having net zero farms.

Altogether, there are many changes needed to develop responsible, sustainable agricultural practices. It is important to keep in mind that many of these changes can also make farming more productive and profitable when done right.

8.2 Land Use

Land is an obvious part of agriculture. Land is used for forests, pasture, and crop production (as well as cities, roads, and parks). As the population has increased, the importance of good stewardship and soil health has received more attention because it is very beneficial to increase crop yields and product quality. Presently, there is a need to use soil more effectively and efficiently in order to have sufficient food for all people.

DOI: 10.1201/9781003396154-8

Forests provide carbon sequestration and wood and paper products. Some forests are managed, while others receive very little attention. Because of the importance of forests for carbon sequestration, improvements in management are recommended in order to enhance production and the benefits of carbon sequestration. Of course, some forests are in areas without easy access, such as in mountain environments or tropical jungles where active management is difficult. It is generally more important to focus on better agricultural practices that can reduce the need to raze even more forests for agricultural purposes.

There are already large plots of land that need to be improved because of poor stewardship. Land with mine tailings can be improved by applying remediation technologies such as phytoremediation (Erickson and Pidlisnyuk, 2021). Adding organic amendments such as compost and/or manure helps to increase organic carbon concentration and microbial numbers in soils. Some land has been impacted by war and has contaminants because of past destructive events. Land mines may be present. It is important to restore these lands to beneficial use not just because it may be morally right, but because these are areas that often have otherwise desirable characteristics (for example, having a good climate and being close to population centers).

Soil health and soil quality are important because crop yields vary with the state of the soil. Physical, chemical, and biological properties of soils can be managed to improve productivity, and soil science education has been beneficial in many parts of the world. About 1/3 of the lands that are used for crop production have soils that could be improved significantly (Erickson and Pidlisnyuk, 2021), leading to increased crop yields that would be very beneficial. The need to increase organic carbon and biological health is a common concern. Soil structure, water-holding capacity, and nutrient availability can be improved by increasing organic matter content in soils. This is not a simple process; managing soil health is a complex topic. Improved soil ecosystems, though, have beneficial value to crop yields, efficient production (leading to better profits), the eventual value of the land itself, and global health.

8.3 Soil Quality

Good soil leads to good crops. What makes for good soil? Briefly, good soil has a lot of organic matter (along with decomposers feeding on it) that breaks down to provide nutrients such as nitrogen, calcium, and phosphorus. Organic carbon in soils is very important for good soil health (Erickson and Pidlisnyuk, 2021). Growing crops, however, uses up these nutrients and leaves the soil depleted. Thus, it is common for agricultural operations to restore or improve soil quality by adding amendments. These supplements can take the form of organic carbon, such as manure, or synthetic ammonia and nitrogen fertilizers.

Adding organic carbon to improve soil health and sequester carbon is recommended at many sites to increase crop yields and decrease carbon dioxide in the atmosphere. There are also many tracts of land that do not have good productive vegetation because of mine tailings, contamination from past uses, degradation because of soil erosion, poor soil management, or other environmental impacts. Restoration processes that lead to soils that can be used for productive agriculture or forestry are desired for many sites. Soil health is thus an important management issue in all agriculture production operations.

Fertilization with organic carbons not only increases the productivity of soils, but it is also a form of carbon sequestration. Because one of the very important processes to include in the path to net zero is carbon sequestration, this should be incentivized as the first-line option for soil supplementation. This could even, theoretically, be used as a pathway for improved carbon sequestration—for example, by using financial incentives to encourage more use of natural carbon fertilization as a form of carbon sequestration (as well as a way to improve food production). Increasing this practice would also help to offset the methane emissions associated with the agricultural sectors due to cattle and other ruminant animals that digest cellulose.

There will nevertheless still continue to be some need for artificial soil supplements—synthetic fertilizers—in agriculture. Even here, though, there are some exciting opportunities. As mentioned earlier, hydrogen production and storage are very important to the transition to green energy. This "green" hydrogen can be used to produce ammonia and nitrogen for fertilizer. Thus, fertilizer can be produced from low-cost solar energy that is used for hydrogen production via electrolysis when electricity use needs to be increased in order to balance supply and demand. Hydrogen can also be used to produce methane from captured carbon dioxide. Ammonia and methane produced from green hydrogen can then be stored and used as needed.

The technology already exists to produce hydrogen by the electrolysis of water using electricity from renewable energy. Solar and wind energy are being used to generate low-cost electricity that can be used to produce hydrogen. Nuclear energy can also generate electricity without combustion emissions. One of the global challenges is to reduce the cost of ammonia production using electricity from carbon-free sources to produce hydrogen (Guizar and Erickson, 2024; Nasser et al., 2022). There are also electrochemical ammonia production processes that can be used to make nitrogen fertilizer (Ghavam et al., 2021; Guizar and Erickson, 2024). Although these electrochemical methods are more expensive, they are of interest because of the importance of reducing greenhouse gas emissions. One of the important current research and development efforts is to reduce the cost of producing hydrogen by the electrolysis of water. Since the cost of renewable electricity is an important part of the cost, developments that reduce the cost of electricity are very valuable to hydrogen production and electricity generation.

In summary, soil quality is a key aspect of farming that has historically been carbon intensive, contributing to greenhouse gas emissions. This does not have to remain the case, though, and can actually be reversed potentially. By moving towards net-zero operations, agriculture could actually take a place alongside forests and the oceans as key parts of the earth important for carbon management.

8.4 Solar Power and Electric Vehicles for Agriculture

Farms and ranches use a lot of equipment. Nearly all of these automobiles, trucks, tractors, and farm implements run on fossil fuels (if they have a motor). Nearly all of these, in principle, can be converted to electric motors. There has been significant progress in developing electric cars and trucks, and many options exist (as described elsewhere in this book). This chapter focuses on the particular opportunities within a farm or ranch context.

One of the things that agricultural settings tend to have in much greater supply than other locations is extra space. Growing up, one of the authors had nearly 40 acres of a family farm to roam over. It is very feasible, therefore, to install basic, ground-based solar panel systems. You can think of it as farming sunlight alongside the other crops. The generated solar power, particularly if supplemented with battery storage, can then be used to power electric trucks, tractors, and other equipment. The cost of electricity that is generated at a farm site has the potential to be less than $0.10/kWh using solar panels and batteries. Importantly, there is a positive interaction between local solar power generation and electric vehicles (Erickson et al., 2016). On-site farm solar power generation for electric farm equipment can reduce both direct and indirect costs for the whole agricultural operation.

There has been less progress in developing electric tractors because of the greater complexity that needs to be addressed. For example, there has been more progress in developing small tractors compared to larger tractors because of the magnitude of the power that is needed and the weight of the batteries (Scolaro et al., 2021; Topal, 2022). Electricity, however, has many developed applications as well as potential future applications on both tractors and associated farm machinery. Hybrid electric tractors, battery electric tractors, and fuel cell electric tractors are included in a comprehensive review by Scolaro et al. (2021).

Battery swap technology also has the potential to be used in agricultural operations to facilitate effective and efficient operations while keeping costs lower. Being able to swap batteries generally allows a user to reduce the size of each battery pack that is used. A large battery electric tractor with 150 kW of power can use more than 500 kWh of energy for a full day of work in some cases. This could be done with a single, very large (and expensive) battery or with multiple smaller batteries that are rotated in and out of use. Consider that, on many other days, that tractor (or other equipment) is not used at all, and those swappable batteries could be utilized in other equipment that is needed at the time. Much like the swappable battery systems that are rapidly dominating the hand power tools sector, battery swap technology can be particularly useful in agricultural settings.

Some immediate details are being worked out right now. There is a need to design and develop battery swap technology that is tailored for use with solar panels on a farm. Some farms do not actually have the space to install solar panel arrays. There are other details that need to be worked out as opportunities. For example, it becomes possible that a farm or ranch with a solar panel array can be connected to the electrical grid with an agreement to both use electricity provided by the utility when needed and get paid to deliver its excess electricity to the grid when it is not needed by the farm. Integration of solar (or wind) power generation can thus even be a source of income stability for agricultural operations.

In summary, there are opportunities to reduce greenhouse gas emissions by adding solar panels to generate electricity for use on farms for on-site homes, crop transportation, and farm operations. Battery swap technology has great potential here in that one can envision being able to charge a battery that can be used for any of the cars, trucks, tractors, combines, or other needs on the farm. The standardization to enable multiple uses for batteries would reduce costs and complexity. Auxiliary batteries may be used in some cases. Multiple smaller (swappable) batteries may be utilized in order to improve efficiency and reduce the weight of the item to be moved.

8.5 Food Waste

In the process of raising animals and growing crops, there are a number of waste products, often referred to as “agricultural by-products.” These by-products include animal manure, food waste, plant waste, wastewater, sewage sludge, and other . . . well, unwanted by-products. Some of these can be fairly directly recycled. Animal manure, as discussed previously, can be used as an organic carbon fertilizer. Some plant waste (e.g., plant bodies after the edible parts have been harvested) can be plowed back into the soil as potential fertilizer for subsequent crops. These are practices that should continue to be encouraged because they are efficient and responsible. Other by-products, though, take more effort.

Many of these other by-products can be handled usefully by utilizing anaerobic digestion. Anaerobic digestion is a process in which microbes are used to produce methane (and carbon dioxide) from various organic wastes. It can, in principle, be used for most types of agricultural waste by-products. The process involves mixed cultures of microorganisms with hydrolysis of polymers, acidogenesis, acetogenesis, and methanogenesis (Neri et al., 2023). The hydrolysis of carbohydrates to sugars, proteins to amino acids, and lipids to fatty acids produces smaller molecular weight organic compounds that are then processed further into smaller organic acids and alcohols. Food wastes have many organic compounds with significant energy content compared to animal manure, which contains organic compounds that remain after processing by the animal. Because of the complexity of the process, it is important to monitor the methane yield and extent of degradation.

The resulting sludge from anaerobic digestion is a good fertilizer product for application to agricultural land. The produced methane is often used to produce energy and/or electricity. One of the benefits of anaerobic digestion is that greenhouse gas emissions are reduced by the anaerobic processing of the organic compounds in these waste products.

In the European Union, there is a government subsidy to encourage anaerobic digestion of animal manure because of the benefits to society in terms of reduced greenhouse gas emissions and improvements to the fertilizer value of the digester solids compared to unprocessed animal manure (Ahlberg-Eliasson et al., 2021). The subsidy has resulted in increased anaerobic digestion of animal manure in Sweden, where the potential annual methane production is capable of generating about 3–6 TWh of electricity (Ahlberg-Eliasson et al., 2021).

There are often benefits associated with anaerobic digestion of mixtures of wastes, including better methane yields (Paranjpe et al., 2023). The need to find good ways to process each of the wastes and the need to have methane from renewable sources are important considerations that combine to result in many anaerobic digesters in many parts of the world. The process is a continuous operation with an organic loading rate that may vary because of variations in the quantity and concentration of wastes that become available to digest.

There is a good review of anaerobic digestion of food waste by Pilarska et al. (2023). Anaerobic digestion is identified in this review as one of the best ways to process food waste. Temperature, pH, C/N ratio, concentrations of fatty acids, organic loading rate, and mean retention time are important variables to monitor and manage. A continuous flow completely mixed digester is often used; however, two-stage systems have been used as well. There are many opportunities to mix food waste with other waste, and benefits have been found in doing this. Pretreatment to reduce particle size is also often beneficial (Pilarska et al., 2023).

8.6 Water

Every living thing needs water, and that, of course, includes both crops and livestock. Water is very important for agriculture. Although there are some places where natural rainfall is sufficient for agricultural activities, irrigation is used in a large proportion of locations. Operations may rely entirely on irrigation or supplement with irrigation when rainfall is short. Irrigation can come from pulling water out of rivers, from wells that tap into small aquifers or large (multi-state) aquifers, or from surface reservoirs and canal systems.

The challenge of how to sustainably manage water needs for agriculture is a huge and pressing issue about which there are already entire books (e.g., Parris, 2010). Most people are now aware that our fresh water resources are shrinking, from the Colorado River that is exhausted before it reaches the ocean, to the massive Ogalala aquifer in the Midwest that is steadily shrinking, to the recurring droughts that overtax California's extensive water reservoir system.

Given that this is both a well-documented and large topic in its own right, we will focus here on an aspect of water in agriculture that gets less attention yet may be more significant in terms of reaching net-zero agricultural practices. That is the topic of agriculture that takes place at the bodies of water (rather than moving water to the places we choose to have agriculture). Food production in and on bodies of water is increasing in importance because of the need for food and the opportunities to increase food production in lakes, rivers, seas, and oceans.

Of the bodies of water that we have on the planet, oceans are, of course, the largest. The net-zero aspect of agriculture includes a focus on the role of oceans because they are important not only for some types of food production but also as a huge player in carbon sequestration.

8.6.1 Oceans

There are opportunities to increase both food production and carbon sequestration in the oceans. First, though, let's give a bit of background. Carbon sequestration in seas and oceans is a fundamentally important part of our global ecosystem, and there are opportunities to increase the amount of carbon dioxide that is removed from the atmosphere and sequestered into water. In particular, phytoplankton (microscopic marine algae) play a massive role in natural carbon sequestration. Phytoplankton provide about 50% of the global primary production of sequestration, which totals about 100 Gt of carbon per year (Kulk et al., 2020). Primary production by phytoplankton is limited by nutrient concentrations in many locations in the ocean. Nitrogen is the limiting nutrient in many locations; however, iron, manganese, and phosphate may be limiting in other locations. Two or more nutrients are limiting in many locations (Browning and Moore, 2023). Phytoplankton production can, therefore, be increased by adding nutrients to the surface waters in most places in the ocean. While there is a good understanding of the importance of nutrient limitation, there is a need for a better understanding of how to add nutrients and manage the increased concentrations of nutrients in the ocean.

One of the opportunities is to greatly increase phytoplankton production in the ocean in order to increase food production and carbon sequestration.

Research is needed to do this in an optimal way. There are many locations in the oceans where nutrient limitation is presently limiting food productivity and carbon sequestration. Research may enable very significant increases in both food production and carbon sequestration at modest costs. There is a need to address many aspects of commercialization of beneficial use of the oceans because systems are not presently in place to lease some areas of the ocean.

The U.S. National Academies of Sciences, Engineering, and Medicine (NASEM) has published a consensus study report on carbon sequestration in the ocean (NASEM, 2022). This report addresses carbon sequestration with an emphasis on carbon dioxide removal from the atmosphere. There are chapters on nutrient fertilization, artificial upwelling and downwelling, seaweed cultivation, recovery of marine ecosystems, ocean alkalinity enhancement, and electrochemical engineering approaches. The report includes a proposed research agenda. The report discusses the need to remove and sequester about 10 Gt of carbon dioxide/year by 2050 and 20 Gt/year by 2100 at a cost of $100/t or less. Knowledge, governance, environmental and social impacts, and monitoring and verification of impacts and processes are identified as important challenges. Estimated costs are in the range of $50–150/t of carbon dioxide; however, there is considerable uncertainty associated with the values. The estimated costs of carbon sequestration may be less if both food production and carbon sequestration are accomplished at the same time with efforts to include both goals in an integrated plan. The scale of the effort may affect costs as well. Nutrient fertilization, seaweed cultivation, and ocean alkalinity enhancement are identified as research priorities.

The complexity of increasing food production and carbon sequestration in the ocean is an important reason for conducting more research through UNESCO and the UN Decade of Ocean Science for Sustainable Development. A recent review on upper ocean biogeochemistry is very helpful (Dai et al., 2023). Subtropical gyres that are regarded as ocean deserts cover about 28% of the ocean surface. This excellent review provides a good understanding of the chemistry and transport phenomena associated with the gyres. Because very large areas of the ocean are not very productive, there are many locations where food production and carbon sequestration can be improved significantly through research and development. There can be a search for better places to implement the ideas and processes that are proposed in the NASEM (2022) report.

The work of Dai et al. (2023) reports that nutrients are available below the surface in many locations and that there are variations in concentrations within the gyres. Additional research is needed to find good places for productive processes and to determine what can be done to enhance productivity at these sites.

There has been some research on public evaluations of alternative technologies for carbon dioxide reduction (Nawaz et al., 2023). Methods to use to make decisions in selecting technologies to use for carbon dioxide reduction are important.

Research on the effectiveness of carbon dioxide reduction processes and the associated risks is beneficial because it is important to achieve the desired results.

Small-scale implementation of a carbon dioxide reduction method to confirm its efficacy and impacts on the ecosystem may be beneficial as part of the effort to determine if a carbon dioxide reduction technology should be implemented at a larger scale. It is important to reduce the risk of unintended side effects, and small-scale implementation of a carbon dioxide reduction technology may be one acceptable and appropriate alternative as part of the UNESCO UN Decade of Ocean Science Research. This proposed research and development strategy may also be used where the goals include both increased food production and carbon dioxide reduction.

One alternative to removing carbon dioxide from the ocean is to use electrochemical processing. The impact of the processing on the environment is small if carbon dioxide is removed from the ocean. A new process has been reported in which carbon dioxide is removed by electrochemical processing at a cost of less than $100/ton of carbon dioxide (Kim et al., 2023). The carbon dioxide removed can be used for some positive purpose or sequestered. This alternative of processing seawater to recover carbon dioxide may be a good topic for further research.

One of the concerns associated with actions to increase food production and carbon sequestration in the oceans is the level of understanding of the impacts of actions on the ecosystem because there may be both positive and negative effects (Maribus, 2024). This new review is an excellent publication on carbon sequestration in the oceans. The oceans are one of the larger sites for carbon sequestration, and much is already happening to make use of the oceans to remove carbon dioxide from the atmosphere. Because of the great need to reduce the concentrations of greenhouse gases in the atmosphere, efforts are being made to find good processes to advance carbon dioxide reduction in the atmosphere by using sites in the ocean for carbon dioxide reduction. The authors advocate moving forward because of the great need to do so, but to consider carefully the risks in the process. The authors state that "climate change is a major environmental disaster," and this is one of the reasons to attempt to find good ways to use carbon dioxide reduction processes to sequester carbon in the oceans (Maribus, 2024). The authors advocate reductions in greenhouse gas emissions should be prioritized.

8.6.2 Sustainable Seafood Production

The management of food production in the open ocean is not simple because of many factors that impact the processes. There are many efforts to increase seafood production in the world, and these efforts are related to several millennium development goals and the UNESCO Decade of Ocean Science. Sustainable mariculture operations that are designed to produce seafood at managed sites in the ocean have the potential to increase seafood harvests significantly.

Employment opportunities in aquaculture and mariculture have a bright future. There is a need for the development of best practices for mariculture operations for a variety of seafood products (Free et al., 2022).

Mariculture has been expanding with many new sites, and the amount of expansion is expected to be limited by product demand. Reduced prices for seafood and increased consumption are expected because of mariculture (Free et al., 2022). There is the potential for a 36–74% increase in seafood production because of mariculture operations (Costello et al., 2020); there is potential for expansion of wild fish production also.

Effective mariculture governance is needed; there are parts of the open ocean far from the land where progress on agreements to establish mariculture operations would be beneficial. One may wish to establish a mariculture operation with a floating system that includes renewable energy production, living quarters, water desalination, waste management, communication system, and global positioning system to maintain a location at a site with good nutrients. Because of the current network of regulations and agreements, there is a need for further efforts to improve governance in some parts of the ocean. The United Nations Convention on the Law of the Sea was approved in 1982, and it entered into force in 1994 (UNCLOS, 1982). It has regulations that are important for the oceans; however, locations close to countries are under regulations developed by the country.

Seaweed production is one of the alternatives that is available; in 2019, 34.7 million tons of seaweed were produced (van der Meer et al., 2023); most of the production was in Asia. Presently, seaweed production for food is modest, and demand is small, with most of the market in Asia. Seaweed production can be increased using mariculture operations; however, there is a need to increase demand, also (van der Meer et al., 2023).

The United Nations Decade of Ocean Science for Sustainable Development (2021–2030) began in 2021. This is a global effort to conduct research related to food production, carbon sequestration, sustainable development, and ocean governance. The research will address marine pollution, ecosystems, biodiversity, food production, economics, ocean-based solutions to climate change, resilience to ocean hazards, ocean observation systems, modeling, capacity to do ocean research, and humanity's relationship with the ocean (Guan et al., 2023). This is an effort to develop the science that is needed for the oceans that are wanted. An implementation plan has been developed and endorsed by the UN General Assembly. The Intergovernmental Oceanographic Commission (IOC) of the UN Educational, Scientific and Cultural Organization (UNESCO) established an executive planning group to provide leadership in the preparation of the Implementation Plan. Research is being conducted in many countries; all disciplines of ocean science are included. Science policy is a priority because of the need for better management of the oceans. The UNESCO website on activities related to the UN Decade of Ocean Science for Sustainable Development (oceandecade.org) includes current information such as white papers that are related to the research that is being conducted.

References

Ahlberg-Eliasson, K., Westerholm, M., Isaksson, S. and Schnurer, A. 2021. Anaerobic digestion of animal manure and influence of organic loading rate and temperature on process performance, microbiology, and methane emission from digestates. Frontiers in Energy Research 9: 740314.

Browning, T.J. and Moore, C.M. 2023. Global analysis of ocean phytoplankton nutrient limitation reveals high prevalence of co-limitation. Nature Communications 14: 5014.

Costello, C., Cao, L., Gelcich, S., et al. 2020. The future of food from the sea. Nature 588: 95–100.

Dai, M., Luo, Y.W., Achterberg, E.P., et al. 2023. Upper ocean biogeochemistry of the oligotrophic north pacific subtropical gyre: From nutrient sources to carbon export. Reviews of Geophysics 61. https://doi.org/10.1029/2022RG000800.

Erickson, L., Robinson, J., Brase, G. and Cutsor, J., Eds. 2016. Solar Powered Infrastructure for Electric Vehicles: A Sustainable Development. CRC Press/Taylor and Francis Group, Boca Raton, FL. https://doi.org/10.1201/9781315370002.

Erickson, L.E. and Pidlisnyuk, V., Eds. 2021. Phytotechnologies with Biomass Production: Sustainable Management of Contaminated Sites. CRC Press, Boca Raton, FL.

Free, C.M., Cabral, R.B., Froehlich, H.E., et al. 2022. Expanding ocean food production under climate change. Nature 605: 490–496.

Ghavam, S., Vahdati, M., Wilson, I.A.G. and Styring, P. 2021. Sustainable ammonia production processes. Frontiers in Energy Research 9: 580808.

Guan, S., Qu, F. and Qiao, F. 2023. United nations decade of ocean science for sustainable development (2021–2030): From innovation of ocean science to science-based governance. Frontiers in Marine Science 9: 1091598.

Guizar Barajas, J. and Erickson, L.E. 2024. Hydrogen and ammonia energy for decarbonization. In Decarbonization Strategies and Drivers to Achieve Carbon Neutrality for Sustainability, M.N.V. Prasad, L.E. Erickson. F.C. Nunes, and B.S. Ramadan, Eds. Elsevier, Amsterdam, 65–83. https://doi.org/10.1016/C2022-0-02538-1

Kim, S., Nitzsche, M.P., Rufer, S.B., et al. 2023. Asymmetric chloride- mediated electrochemical process for carbon dioxide removal from oceanwater. Energy and Environmental Science 16: 2030.

Kulk, G., Platt, T., Dingle, J., et al. 2020. Primary production, an index of climate change in the ocean: Satellite-based estimates over two decades. Remote Sensing 12: 826. https://doi.org10.3390/rs12050826.

Maribus. 2024. World Ocean Review 8: The Ocean—A Climate Champion? How to Boost Marine Carbon Dioxide Uptake. Maribus gGmbH, Hamburg. https://worldoceanreview.com/en/wor-8/.

NASEM. 2022. A Research Strategy for Ocean-based Carbon Dioxide Removal and Sequestration. The National Academies Press, Washington, DC. https://doi.org/10.17226/26278.

Nasser, M., Megahed, T.F., Ookawara, S. and Hassan, H. 2022. A review of water electrolysis-based systems for hydrogen production using hybrid/solar/wind energy systems. Environmental Science and Pollution Research 29: 86994–87018.

Nawaz, S., St-Laurent, G.P. and Satterfield, T. 2023. Public evaluations of four approaches to ocean-based carbon dioxide removal. Climate Policy 23: 379–394.

Neri, A., Bernardi, B., Zimbalatti, G. and Benalia, S. 2023. An overview of anaerobic digestion of agricultural by-products and food waste for biomethane production. Energies 16: 6851.

Paranjpe, A., Saxena, S. and Jain, P. 2023. A review on performance improvement of anaerobic digestion using co-digestion of food waste and sewage sludge. Journal of Environmental Management 338: 117733.

Parris, K. 2010. Sustainable Management of Water Resources in Agriculture. OECD, Paris, France. OECD is the Organisation for Economic Cooperation and Development.

Pilarska, A.A., Kulupa, T., Kubiak, A., et al. 2023. Anaerobic digestion of food waste—A short review. Energies 16: 5742.

Scolaro, E., Beligoj, M., Estevez, M.P., et al. 2021. Electrification of agricultural machinery: A review. IEEE Access 9: 164520–164540.

Topal, O. 2022. A literature review on electric tractors and assessment of using for Turkey. Journal of Agricultural Machinery Science 18: 114–125.

UNCLOS. 1982. United Nations Convention on the Law of the Sea. United Nations, New York. www.un.org

Van der Meer, J., Callier, M., Fabi, G., et al. 2023. The carrying capacity of the seas and oceans for future sustainable food production: Current scientific knowledge gaps. Food and Energy Security. https//doi.org/10.1002/fes3.464.

9

The Complex Role of Hydrogen

9.1 Introduction

Hydrogen might well turn out to be a crucial piece of the journey to net zero. It seems to be not appreciated for its potential, however, by many people. Perhaps this is because hydrogen production is less visible than solar panels and wind turbines. Perhaps this is because the production and uses of hydrogen are more complicated. Nevertheless, hydrogen is already very important in many ways. It is in water, it is in a wide array of chemical compounds, it is part of most of our foods, and hydrogen is part of energy sources such as natural gas, gasoline, and ethanol.

Our focus here, though, is on pure hydrogen. Pure hydrogen is a gas that can be used as a fuel via combustion or to power a fuel cell to produce electricity. Hydrogen can also be combined with nitrogen to produce ammonia, which is an important compound for agricultural fertilizers, as a form of fuel, and is useful in many other ways. This chapter ends with some consideration of the key issues associated with hydrogen as a fuel, such as its transportation, storage, use, and safety.

9.2 Production of Hydrogen

Hydrogen can be produced via several different means. Hydrogen produced in a traditional way from methane in natural gas is sometimes distinguished by calling it *gray* hydrogen. Hydrogen can also be produced by electrolysis of water, and when electricity from renewable sources such as wind and solar is used to produce hydrogen in this way, it has been termed *green* hydrogen. Additionally, *blue* hydrogen is the term for hydrogen derived from methane if the carbon dioxide byproduct is sequestered. In many ways, the key challenge for transitioning from gray to green hydrogen is simply cost. In 2023, the reported average costs are \$2.13/kg for gray hydrogen, \$3.10/kg for blue hydrogen, and \$6.40/kg for green hydrogen (Schelling, 2023).

DOI: 10.1201/9781003396154-9

Hydrogen is an essential element for many industrial processes. It is used for petroleum refining, ammonia production, fuel cells, and other purposes (Massarweh et al., 2023). In 2018, the global use of hydrogen was about 73.9 million tons/year, with petroleum refining and ammonia production as the two largest uses (Massarweh et al., 2023). Recall that carbon dioxide is produced in conventional processes to produce gray hydrogen, so these are industries that contribute to climate change rather than moving towards net zero. Green hydrogen, therefore, can have an impact both in terms of meeting energy demands without increasing emissions and in terms of replacing pollution-generating energy processes.

9.2.1 Production of Green Hydrogen

The production of hydrogen from renewable resources, in which there are no greenhouse gas emissions, is an important goal associated with the ongoing efforts to reduce emissions overall. Because a key issue impeding this goal is the cost of green hydrogen, the U.S. Government started an effort in 2024 to reduce the cost of green hydrogen production significantly, and incentives are currently available (Guizar and Erickson, 2024).

The U.S. Inflation Reduction Act of 2022 (H.R. 5376, 2022), in particular, provides incentives for the production of green hydrogen. One of the important resources for producing green hydrogen is inexpensive electricity from renewable sources such as wind and solar energy. As part of COP 28 in 2023, there was an agreement to triple renewable power capacity by 2030 in order to reduce greenhouse gas emissions. Wind power, solar power, and geothermal power are all part of that process, but the production of green hydrogen could well be crucial to meeting that goal. We can greatly increase the production of green hydrogen by producing inexpensive renewable electricity for the grid and then using that (excess) energy for hydrogen production via the electrolysis of water. That green hydrogen can then replace gray hydrogen, both increasing renewable power capacity and simultaneously decreasing non-renewable power dependency. In order to produce hydrogen from water at $2.00/kg, it is desirable to have electricity from wind and solar that costs less than $0.03/kWh. As discussed earlier, batteries can be used to store renewable energy for later use in the grid, for swapping, and for charging EVs. Green hydrogen can also be thought of in this context as a way of storing energy for later use. These are all important for demand management because the storage systems can be charged up when renewable electricity is available but not needed for immediate uses.

We are already starting to see some progress in reducing the cost of green hydrogen (Zainal et al., 2024). The best results have been obtained with the solid oxide electrolysis cell. The U.S. goal is to reduce the cost of green hydrogen production to $1/kg of hydrogen by 2030 (Zainal et al., 2024). A feasible goal would be the production of 100 million kg/year of green hydrogen by 2040, and that would be very beneficial for the effort to reduce greenhouse gas emissions.

How could this happen? Consider, for instance, the case of a coal-fired electricity generating plant run by Evergy at a site near Lawrence, Kansas. This coal-fired power plant is expensive to operate, and the cost of electricity production with wind and solar is much less in Kansas. It exists and is still used because it is part of the legacy infrastructure and cannot be replaced yet. A site such as this may be a good place to produce hydrogen because there are transmission lines, water lines, land for electrolysis equipment, and space for hydrogen storage. Interstate 70 is nearby to locate a site for hydrogen fueling, battery swapping, and EV charging. Solar panels could be installed on-site or nearby to produce electricity. Hydrogen produced and stored at the site could be used to produce electricity when that is needed. Demand-based energy could be provided using fuel cells or by feeding hydrogen into the coal-burning power plant.

Yet another important use for hydrogen is the production of ammonia, which is used as fertilizer. Continuing the previous example, the hydrogen produced at the Lawrence site could be used to make ammonia for fertilizer use. It is already the case that ammonia fertilizer has historically been produced in Lawrence, Kansas, and there is a local market for it in eastern Kansas.

9.2.2 Production of Blue Hydrogen

Blue hydrogen production includes the capture and storage of the carbon dioxide which results as a byproduct of the conventional production of hydrogen. The conventional production method most commonly used is the steam methane reforming process, in which natural gas is used (Massarweh et al., 2023). Gray hydrogen is produced using the steam methane reforming process, and gray hydrogen becomes blue hydrogen after the carbon capture and storage element is added. Thus, the cost (and complexity) of blue hydrogen will always be greater than that of gray hydrogen, and the cost of gray hydrogen depends on the cost of natural gas if the steam methane reforming process is used. A recent review (Massarweh et al., 2023) used a starting price of $0.94/kg for gray hydrogen gas. After carbon capture and storage, the cost of blue hydrogen was $1.50/kg.

Why then talk about blue hydrogen? First, it can be a useful transition stage for moving partially towards a more responsible form of hydrogen. The carbon sequestration technologies developed for blue hydrogen may eventually even be useful in other contexts. Second, there will always be a need to appropriately account for the byproducts, wastes, and pollution from energy generation. The fact that gray hydrogen production dumps carbon dioxide into the atmosphere has to be considered, just as one would expect accountability for someone dumping a pile of trash in your front yard. Whether the accountability is in the form of regulations that limit the amount of pollution from gray hydrogen production or carbon exchange marketplaces that assign appropriate costs to the environment of that pollution, there needs to be accountability.

9.3 Renewable Fuels

Progress in green hydrogen processing is one of the global challenges that is being addressed in many countries. Various places are working on green hydrogen, ammonia production, methane from anaerobic digestion, ethanol from biological sources, and other fuels from biomass processing. The annual production is increasing, and the cost of production is decreasing as research continues. There are also efforts to capture methane emissions and use the methane for beneficial purposes. Another area of research is to transition to the production of green ammonia such that there are no greenhouse gas emissions associated with the process because green hydrogen is used. There is interest in using ammonia as a fuel that has no carbon emissions, and this is a subject of research. Some notable features of this approach are that the storage and transport of ammonia are well-developed, and the cost is less than that of hydrogen.

Using ammonia as a fuel can address many important issues, including safety and economic considerations (Valera-Medina et al., 2021). Ammonia is a very good fuel for storing energy because storage costs are low, the fuel does not contain carbon, and there is a history of ammonia storage because it is used as a fertilizer. The production of green ammonia is, therefore, an area of current research.

As we saw with gray versus green hydrogen, a key consideration is how green ammonia compares to traditional (gray ammonia) production methods. The current production method for ammonia is using the Haber-Bosch process, and greenhouse gas byproducts are about 146 million tons/year. Because of the large amount of carbon dioxide emissions associated with gray ammonia production, the transition to green hydrogen and green ammonia to reduce emissions of carbon dioxide is important. Global ammonia production is the second largest contributor to those emissions in amount by mass per year (Bertagni et al., 2023). There is even ongoing research to develop new processes that directly produce ammonia by electrochemical conversion of water and nitrogen to ammonia (Bertagni et al., 2023). For an excellent review of this area, see Valera-Medina et al. (2021).

Using ammonia as a combustion fuel has some potential, but there are also some difficulties to overcome. The combustion of pure ammonia has an issue with ignition delay time. Mixing hydrogen or natural gas with ammonia helps with this, and the ignition delay time is reduced (Valera-Medina et al., 2021). Hydrogen, as compared to natural gas, is a good substance to add to improve the combustion of ammonia because it does not have carbon emissions. Another important concern associated with combustion of ammonia, though, is the production and emission of nitrous oxide (N_2O), which is a particularly potent greenhouse gas with a global warming potential 265 times a comparable amount of carbon dioxide (Bertagni et al., 2023). Nitrous oxide emissions can be reduced by good management of the combustion process, but this must be recognized as an issue.

Lastly, there is a significant history of ethanol as a possible renewable fuel. Many people are at least familiar with the existence of ethanol as a fuel, at least in the context of ethanol added to gasoline for internal combustion vehicles (for example, the E-85 blend at filling stations). This ethanol largely comes from the fermentation of glucose by the processing of farm crops such as corn. For historical, practical, and political reasons, the use of ethanol as a gasoline additive has become commonplace in the United States. How good is it in terms of helping us move to net zero? Ethanol has recently been compared to electricity produced using solar PV, and ethanol from corn requires 85 times more land compared to solar panels to produce electricity to travel the same distance (Mathewson and Bosch, 2023). That means that there are locations, such as southwest Kansas, where irrigation is used to grow crops to produce ethanol that could be converted to solar PV and would produce more renewable electricity with reduced greenhouse gas emissions.

9.4 Storage and Transportation of Renewable Fuels

The storage and transportation of renewable fuels are important, and there are costs associated with both. The length of time associated with storage is also an important consideration. Both winter heating and summer air conditioning have impacts on fuel use and, therefore, storage needs. Ammonia can be generated at times that it is advantageous to do so, such as when surplus renewable energy is available, but then it needs to be stored so that it can be applied as a fertilizer at the appropriate times. Fortunately, there are several options, and a good review related to hydrogen transportation and storage was by Yang et al. (2023). Hydrogen can be stored as a gas under pressure, as a liquid at low temperatures, or even in a cryo-compressed supercritical state.

Hydrogen can be used in fuel cells and, to the extent that fuel cell electric vehicles (FCEVs) become more common, there will be a need for stations that sell hydrogen. Hydrogen can be stored as a gas at these stations. A typical fuel tank for an FCEV may be about 120 liters in capacity with the ability to have a pressure of about 700 bar (70 MPa) to hold about 5 kg of hydrogen. With current FCEV technology, this will provide a range of about 500 km; hydrogen contains a large amount of energy per unit of mass. The maximum efficiency of a fuel cell is about 60% (Hassan et al., 2023). That means that FCEVs are much more efficient than cars with internal combustion engines, although FCEVs are currently less efficient than BEVs.

Storage tanks for hydrogen are important because of the storage conditions and the small size of hydrogen molecules. There is a good review of the types of tanks and manufacturing methods in Cheng et al. (2024). They review five types of hydrogen tanks for storage. Tanks for hydrogen storage in FCEVs are currently made with a metallic liner of aluminum alloys and wrapped with composite. These tanks are produced without welding (Cheng et al., 2024).

In summary, hydrogen can play a critical role in the transition to a net zero environment. The production, storage, and use of hydrogen can be complex, and its potential is probably underappreciated. But the flexibility of hydrogen in all its roles makes it an important part of the renewable energy mix. Furthermore, to the extent that green options can replace the use of gray hydrogen and gray ammonia, there are effective and important reductions in greenhouse gas emissions.

References

Bertagni, M.B., Socolow, R.H., Martirez, J.M.P., et al. 2023. Minimizing the impacts of the ammonia economy on the nitrogen cycle and climate. PNAS Earth, Atmospheric, and Planetary Sciences 120(46). https://doi.org/10.1073/pnas.2311728120.

Cheng, Q., Zhang, R., Shi, Z. and Lin, J. 2024. Review of common hydrogen storage tanks and current manufacturing methods for aluminum alloy tank liners. International Journal of Lightweight Materials and Manufacture 7: 269–284.

Guizar, J.D. and Erickson, L.E. 2024. Hydrogen and ammonia energy for decarbonization. In Decarbonization Strategies and Drivers to Achieve Carbon Neutrality for Sustainability, M.N.V. Prasad, L.E. Erickson, F.C. Nunes, and B.S. Ramadan, Eds. Elsevier, Amsterdam, 65–83.

Hassan, Q., Azzawi, I.D.J., Sameen, A.Z. and Salman, H.M. 2023. Hydrogen fuel cell vehicles: Opportunities and challenges. Sustainability 15: 11501.

H.R. 5376. 2022. Inflation Reduction Act of 2022. Public Law No. 117–169 (08/16/2022). Library of Congress. congress.gov.

Massarweh, O., Al-khuzaei, M., Al-Shafi, M., et al. 2023. Blue hydrogen production from natural gas reservoirs: A review of application and feasibility. Journal of CO2 Utilization 70: 102438.

Mathewson, P. and Bosch, N. 2023. Corn ethanol vs solar land use comparison. Clean Wisconsin. cleanwisconsin.org/

Schelling, K. 2023. Green hydrogen to undercut gray sibling by end of decade. Bloomberg NEF Newsletter, August 9. bnef.com

Valera-Medina, A., Amer-Hatem, F., Azad, A.K., et al. 2021. Review on ammonia as a potential fuel: From synthesis to economics. Energy & Fuels 35: 6964–7029.

Yang, M., Hunger, R., Berrettoni, S., et al. 2023. A review of hydrogen storage and transport technologies. Clean Energy 7: 190–216.

Zainal, B.S., Ker, P.J., Mohamed, H., et al. 2024. Recent advancement and assessment of green hydrogen production technologies. Renewable and Sustainable Energy Reviews 189: 113941.

10

Economics and Policies for Sustainable Energy

Moving from our present global systems to net-zero systems in energy production, transportation, living/working arrangements, and agriculture can be a daunting prospect. They all can be made easier to understand—if not implemented—by thinking of them in terms of basic market economics. That is, what are the costs and benefits of taking one course of action (e.g., heating a home with coal), and what are the costs and benefits of taking another course of action (e.g., heating a home with natural gas)? If there are multiple possible courses of action, the costs and benefits of each can be worked out, but in the end, one usually has to decide on which option is the best one. "Best" here is often simply the one with the best ratio of benefits to costs. Economists assume that, generally, people are rational cost/benefit analysts, and they will go with the best option for them. A transition to a better sustainable energy option will occur whenever the benefit/cost ratio of the renewable energy technology is better than the alternative options.

Sometimes, there are factors that people want to include in these decisions about one option versus another, but they are not clear cost/benefit factors. Such considerations can be inserted into the basic analysis here, implementing relevant policies. For example, a policy that subsidizes fossil fuels (exploration, development, tax breaks, etc.) will delay any movement away from fossil fuels and to other energy sources; the policy either absorbs some of the usual costs of the option or enhances the benefits (or both). One can think of this as metaphorically putting a thumb on the scale as people weigh the costs and benefits of their options. Moving to sustainable energy can be accelerated by policies that subsidize renewable energy generation, electric vehicles, energy efficiency in buildings, and sustainable agriculture. Policies can be a way to include long-term costs of actions (e.g., effects of climate change) into the immediate, day-to-day evaluations and decisions of individuals. Policies are basically reflections of values, above and beyond just economic costs and benefits.

10.1 Economics of Sustainable Energy

Economics is a very important topic because it is impacting efforts to reach net zero greenhouse gas emissions in myriad ways. Most people consider prices in

DOI: 10.1201/9781003396154-10

making decisions, and the price of sustainable energy practices has been steadily dropping. The costs of electricity generation for solar and wind processes have rapidly decreased. The cost of batteries for electric vehicles and for storing renewable electricity has decreased significantly since 2010. The cost of producing hydrogen via electrolysis of water is still relatively high, but there is ongoing research on ways to decrease it (recall that solar and wind energy used to be very expensive at one time). The technology for producing ammonia (for use as nitrogen fertilizer) using renewable electricity has been developed, but the costs there also need to be reduced. Even in specific consumer areas, prices have consistently been dropping. For instance, electric bicycles have been an area with huge perceived potential, and many people have worked on developing and marketing electric bicycles as affordable and meeting the needs of many people (particularly in developing parts of the world). One of the key factors, however, that allows the adoption of electric bicycles to take off is when there has been progress in reducing the cost of production and the selling prices of electric vehicles.

Thus, economic concerns and prices are impacting the transition to renewable electricity, battery electric vehicles, and net zero greenhouse gas emissions for homes, public buildings, transportation, and industrial companies.

10.1.1 Power Generation and Storage

The efforts to reduce greenhouse gas emissions, transition to renewable energy to generate electricity, and electrify transportation have been in progress for more than 20 years. Costs and prices have impacted progress, and there has been better success where new developments have enabled prices to be reduced. The generation of electricity using solar and wind is a good example of progress because electricity costs for these two renewable generation processes in 2024 are about $0.03/kWh in some locations. There are many locations where these alternatives are the lowest cost compared to natural gas, coal, and nuclear energy. Because of the low cost of wind and solar generation of electricity, 86% of electric power capacity additions in 2023 were with renewable power; of the 473 GW of new renewable capacity, more than 50% was in China. In 2023, solar growth accounted for 73% of the new renewable installations (IRENA, 2024).

One important issue that must be addressed is the ability to meet the demand for electricity at all times reliably. Since the generation of electricity using wind and solar is not dispatchable, it is necessary to find other ways to meet demands. Batteries currently appear to be a likely solution to this issue of being able to store energy that can be used later (that is, being dispatchable). There has been good progress in the development of batteries for energy storage which are being used as part of the electrical grid and for EVs.

10.1.2 Sustainable Transportation and Structures

There has been good progress in the electrification of transportation because of a significant reduction in costs to manufacture batteries. Tesla, Inc. has a history of development and progress of more than 20 years since it was incorporated in 2003. As of 2024, many battery electric vehicles and plug-in hybrid electric vehicles are competitive and selling well in many countries (Woody et al., 2024). The smaller battery electric vehicles do very well compared to other alternatives based on total cost analysis. Battery electric vehicles have low maintenance costs. Because of efforts to improve efficiency and other developments, prices for battery electric vehicles have been reduced, and they are being sold at relatively low prices in some countries (Erickson, 2024). The expectation is that global sales of EVs will reach 20% of the automobile new car market in 2024, with about 16.7 million sales and a growth rate of 22% (McKerracher, 2024).

Many owners are able to charge their EVs at home or where they work at very reasonable prices. In some cases, incentives are provided to charge EVs in order to manage the demand for renewable electricity (Erickson and Ma, 2021). Because of the low prices of solar panels, some people have installed solar panels on their homes or businesses to generate electricity, and some of the electricity is used to charge their battery electric vehicles. Some parking lots have been covered with solar panels to provide shaded parking and opportunities to charge EVs at low prices while shopping or at work. The prices for charging have been reviewed, and charging at home at level one is often among the lower-cost options (Borlaug et al., 2020; Lanz et al., 2022). Fast charging is often more expensive, with prices more than twice those of charging at home.

Battery swapping has been developed in more limited areas to provide a very rapid battery alternative for fleets of buses and taxis. This is a very good alternative because depleted batteries can be charged when low-cost electricity is available as part of the demand management associated with renewable energy production (Erickson and Ma, 2021). Battery swapping has been developed and used successfully in China, and the economics of this have been investigated and reported (Wu et al., 2022). The costs associated with swapping cannot be directly compared to battery charge stations because the charging of the battery provided with the swapping is done by the company that does the swapping. The charging of batteries can be carried out at times that are beneficial for the grid and for demand management of solar and wind generation of electricity (Revankar and Kalkhambkar, 2021).

Another recent development in China has been reported by Contemporary Amperex Technology Co. Ltd (CATL). CATL is introducing modular batteries that can be added as needed to provide supplemental energy for cars and trucks (Alanazi, 2023; Hampel, 2023; Qiang et al., 2023). The module for BEV cars is for about 40kWh and 200km, while the module for trucks is for about 170kWh. The concept of having multiple batteries or supplemental batteries that can be added when needed for a longer trip has merit. The benefit is that these modular batteries can be installed on many different EVs and also used for grid electricity when needed.

The development of battery charging together with battery swapping is very beneficial because both are needed. There are benefits to having an EV that can make use of all of the options that are available.

10.1.3 The Economics of Doing Nothing

There is an inertia in human nature; we will tend to continue doing whatever we have done in the past. A good economic analysis of options, though, has to consider the costs and benefits of that inertia, of doing nothing different from the past. We know that our past and current patterns of behavior lead to systemic and global climate changes; to deny that fact at this point requires so many perversions of information that it can be ignored as a valid position by any objective person. The real question of importance (economically) is what the costs of climate change are and what they will be going forward.

The costs associated with climate change have been increasing, and they are significant. In 2023, the United States experienced 28 disasters associated with climate change. This is a record number of separate events that each cost more than one billion dollars (Smith, 2024). In this report from January 8, 2024, the total cost for 2023 had reached $92.9 billion, and additional costs may be added. The prior record was 22 events in 2020. There were 492 fatalities in 2023. The cost for the period 2019–2023 has reached $603.1 billion. Drought and heat in southern and midwestern states cost $14.5 billion, with 247 associated deaths. The island of Maui in Hawaii experienced 100 deaths and $5.6 billion in costs because of wildfire damages.

The global costs of disasters associated with climate change have been increasing. A sevenfold increase in the number of global disasters has been reported for the period 1970–2019 by the World Meteorological Organization (Newman and Noy, 2023). One of the ongoing areas of research is to estimate the impact of climate change on the costs of each disaster. Extreme event attribution research is being conducted, and the estimate is that $143 billion of the global annual cost of disasters is due to climate change.

The social cost of carbon dioxide has been used to communicate the estimated impact of adding carbon dioxide to the atmosphere. It is another way of bringing the cost of climate change into policy decisions. One recent estimate is $185/ton of carbon dioxide (Rennert et al., 2022). This may be compared to the value of $51/ton that is used by the U.S. government to communicate the economic impact of climate change. There is a good literature review on the economic impact of climate change, including the social cost of carbon dioxide, in Tol (2024).

One of the options for mitigating the costs of climate change is to increase the amount of carbon sequestration that is occurring. There is already a lot of carbon sequestration occurring all over the world in the form of plants using carbon during photosynthesis. We can enhance these to some extent by, for example, having better forest management and healthier stocks of phytoplankton in the oceans. Often, though, this option is about developing and using technologies to accomplish further carbon sequestration artificially.

This is both a massively ambitious task and a relatively small task at the same time. Consider that the global quantity of carbon dioxide released into the atmosphere was about 37 Gt in 2022 (37 billion metric tons). If carbon dioxide can be removed from the atmosphere at a cost of $50/t, it would cost $1.85 trillion/year to remove the 37 Gt that is added each year, which is about 1.85% of the global gross domestic product of about $100 trillion/yr.

Carbon capture and sequestration processes to remove carbon dioxide during industrial production have been developed and are being implemented. Costs vary from $5–66/ton of carbon dioxide (Hughes and Zoelle, 2022; Pilorge et al., 2020). There are federal tax credits for qualifying facilities in the United States. In Canada, costs from CAD 27–48/ton of carbon dioxide have been reported for concentrated streams and CAD 50–150/ton for diluted gas streams (Sievert et al., 2023).

Air can be processed to remove carbon dioxide from the atmosphere, but it costs about $500–1000/ton with the industrial processes that are presently used (Gelles, 2024). Carbon dioxide is present in much larger concentrations in the oceans as carbon dioxide and bicarbonate and carbonate ions. The concentration increases with pH; it is about 60 times larger and equal to about 0.001 molar in ocean water. Chemical separation processes can be used to process ocean water and remove carbon dioxide at a cost of about $50–100/ton (Kim et al., 2023; Service, 2024). This cost may be reduced through further research and development.

Carbon sequestration technology and cost are two of many factors that need to be confronted if one considers not moving (or not moving quickly enough) to a net zero world. There are other forms of contamination that need to be dealt with. Many locations around the world are contaminated from mining, military activities, pesticide contamination, past industrial use, and other uses where ecosystem restoration and productive use would be beneficial. This need has been recognized, and the United Nations has established the UN Decade of Ecosystem Restoration 2021–2030. Forest establishment may be a good alternative in some cases. Greater efforts are needed to improve soil quality and establish productive uses at many sites in all countries of the world. Phytotechnologies have been developed to improve soil quality for many contamination problems (Erickson and Pidlisnyuk, 2021).

10.2 Policies for Sustainable Energy

The economics of different courses of action shift over time. Certain costs will decrease with better technologies or will increase because of supply chain pressures. Benefits (per unit of cost) will similarly decrease or increase also with changes in situations. This is neither inherently good nor bad, but of course, in the wider world of consequences and preferences, they can be evaluated in terms of desired or undesired consequences.

For example, broad recognition of the consequences of greenhouse gas emissions and climate change has led most people to conclude that these are bad things (for us, for future generations, for global quality of life, etc.). This conclusion leads to efforts to elucidate just how we can reduce greenhouse gas emissions and reach net zero greenhouse gas emissions. Formally, these efforts have resulted in the creation of the United Nations Framework Convention on Climate Change (UNFCCC) and the Paris Agreement on Climate Change, which has the goals of reaching net zero greenhouse gas emissions by 2050 and keeping the increase in average global temperature to less than 2 degrees Celsius (Erickson and Brase, 2020; UNFCCC, 2015).

These are policies in the form of position statements—providing guidance about what should be done or what needs to be done. They do not, however, tend to get into specifics. First of all, it is exceptionally difficult to be specific when confronting a global problem. Specific details about courses of action to address a problem (what actions are possible, allowed, will be adopted by the community, and so on) usually depend on the particular context in which one is tackling that problem. Second of all, the United Nations is not an entity with a tremendous amount of power to enforce rules, regulations, or policies. They often function as a source of general guidance, with individual countries deciding how best to heed those guidelines.

What is usually needed in order to effect change are policies with "teeth," policies that substantively change the costs and benefits of different actions in order to alter behaviors. That is, policy can shape economics. In democratic societies, policy can generally nudge people and institutions in certain directions but is rarely drastic enough to force rapid changes. Often, policies do their nudging via increased taxes on some particular sector, or tax rebates, or tax subsidies, and so on. If people and companies in that sector do a particular action (or don't do some particular action), there is a better benefit/cost ratio.

Policies themselves are neither good nor bad; they are reflections of the values we decide to impose on ourselves and on our society. Thus, there are many different specific sets of policies about the topics covered in this book, depending on what society you happen to live in. The following sections will focus on the policies of the United States, both because it is a major world power (and greenhouse gas emitter) and because we are familiar with these policies.

10.2.1 Power Generation and Storage

Consistent with the UN guidelines, the U.S. government has developed plans and approved incentives to encourage the reduction of greenhouse gas emissions across a timeline of many years. On August 16, 2022, the Inflation

Reduction Act of 2022 became law (H.R. 5376, 2022). This legislation is an important policy act with many incentives to encourage actions to reduce greenhouse gas emissions. In particular, Subtitle D on energy security has several provisions. It modifies and extends tax credits for electricity production from wind and solar systems. Energy storage is included, as well as carbon oxide sequestration. There are tax credits for biodiesel, sustainable aviation fuel, and qualified clean hydrogen.

One very significant incentive within the Inflation Reduction Act of 2022 is the tax credit for the production of hydrogen via electrolysis of water. Hydrogen is needed for fuel use, the generation of electricity, and the production of ammonia, and it has many uses in the production of other products.

The Inflation Reduction Act also provides tax credits for residential clean energy, energy efficiency expenditures, and qualified battery storage expenditures. Presently, the combustion of natural gas and other fuels is used to heat air and water in many buildings. There are now tax deductions for the improvement in efficiency and greenhouse gas emission reductions in commercial buildings. There is also a $5000 tax credit for certified new zero-energy homes (H.R. 5376, 2022).

10.2.2 Sustainable Transportation and Structures

The Inflation Reduction Act of 2022 continues the use of tax credits to encourage the purchase of new (and now used) plug-in EVs for personal and commercial use. There is a 30% tax credit for investments in qualifying energy equipment manufacturing projects that produce renewable electricity equipment. There is a tax credit for investment in facilities for energy storage operations and for domestic clean fuel production.

The Inflation Reduction Act provides funding to USDA for incentives to extend the goals of the act to rural communities and organizations such as electric cooperatives. The U.S. Department of Energy receives funding to provide energy rebates, energy efficiency programs, and loan programs. Loans from DoE are available for energy infrastructure projects. Grants are available for the domestic production of efficient electric vehicles (H.R. 5376, 2022). Loans for energy resource development are provided to Indian tribes, and loans are provided for electric transmission lines. Funding is also provided for clean energy demonstration projects to deploy advanced industrial technology. To improve air quality, funding for zero-emission vehicles is provided for school buses, and grants to port authorities are made to reduce pollution. Finally, the act addresses the methane emissions reduction program and provides financial incentives in support of the goals of the program (H.R. 5376, 2022).

One important alternative is carbon sequestration associated with forest growth, forest management, wildfire prevention, and forest restoration. The USDA Act provides funding to the National Forest System for activities that increase carbon sequestration. Grants are provided to states, Indian tribes, and local governments for tree planting.

10.2.3 U.S. Implementation Plans

The United States Government has developed five goals as part of the Federal Sustainability Plan (Biden, 2021). These are:

1. All electricity for the federal government will be locally supplied carbon-free electricity (CFE) by 2030.
2. By 2035, all vehicle acquisitions will be zero-emission vehicles.
3. By 2050, all federal procurements will have net-zero emissions.
4. By 2045, all federal buildings will have net-zero emissions.
5. By 2050, all federal operations will have net-zero emissions.

In the process of moving forward, this will include an effort to reduce greenhouse gas emissions by 65% by 2030 and achieve net-zero emissions by 2050.

Implementation of all this—multiple, simultaneous efforts to shift towards renewable energy and stop increases in global climate change—is a tremendous logistical challenge. There are, therefore, multiple preparations ongoing to help these goals happen. Public and private partnerships with states, Indian tribes, private companies, and consumers are being developed to accelerate progress (Biden, 2021). The Council for Environmental Quality is charged with providing leadership on the development of performance standards for net-zero federal buildings with respect to efficiency and decarbonization. A working group of federal employees and a task force that has been educated to purchase clean products will lead the transition to net-zero emissions procurement and the modification of regulations to incorporate the social cost of greenhouse gas emissions.

Within the federal government itself, a number of implementation steps are built into these goals and are now being done. The federal government will modernize its own policies, infrastructure, operations, and programs with the goal of reaching net-zero greenhouse gas emissions by 2050. Climate risk management is to be included in the implementation of this federal sustainability plan (Biden, 2021). There are already plans developed on how to move toward net-zero buildings in the United States, and a national publication is now available (DoE, 2024). There are about 130 million buildings in the U.S., and about one-third of greenhouse gas emissions are from combustion produced in these buildings at an annual cost of about $370 billion (DoE, 2024). There is a need to transition to net-zero greenhouse gas emissions in all of these buildings by improving energy efficiency, electrifying space and water heating, and increasing the emphasis on community resilience. Many buildings will need to be modified between 2024 and 2050 to reduce greenhouse gas emissions. The plan is to transition to make building renovations that achieve operations with zero greenhouse gas emissions by 2030 and new building operations that feature zero greenhouse gas emissions. In the process of transitioning to heat pumps that use electrical power, efficiency will be improved, and costs will be reduced.

Along with addressing the built infrastructure of the US federal government, the plan includes education and training of the federal workforce on the goals, plans, and transition processes, as well as the expected benefits of the changes that are being implemented to embed sustainability into the culture of the workforce. Environmental justice and equity are important goals in the United States that are already being implemented within the federal workforce and within many organizations, and they are to be included in all aspects of the federal sustainability plan (Biden, 2021).

10.3 The Intersection of Economics and Policy

The previous sections of this chapter are simple reviews of the economics (costs and benefits) of different sectors and the policy implications (things that can be done to alter those costs and benefits). The real world is complex, though, and there are a number of other factors that do, in real terms, shape how people behave (see also the next chapter on human decision-making). This final section describes some of these factors that tend to come at the intersection of basic economics, policy, and different perspectives.

First of all, we noted before that there is a certain amount of inertia in human nature. Some of this may be due to simply being more comfortable with the current state of affairs and not wanting to adjust or learn anything new. There are other forces maintaining this inertia, though. People sometimes have significant prior investments in the existing system (these investments can be financial, philosophical, or even emotional). Along with this inertial force, there are uncertainties on the other side, such as possible transition costs and uncertainty about the reliability of newer technologies.

Perhaps even more importantly, we need to recognize that nearly all economic analyses of benefits and costs will change whenever one shifts the identity of the person doing those analyses. In other words, it is important to ask, "Whose costs?" and "Whose benefits?" These costs and benefits are not necessarily aligned across individual consumers, communities, countries, and corporations. For any activity, the benefits are not necessarily realized equally across the different actors involved, nor are the costs borne equally across them.

On a very large scale, this is illustrated by thinking about the benefits of fossil fuel usage and the costs of associated greenhouse gas emissions. The benefits of fossil fuel usage go to oil producers, distributors, and end users. The costs of emissions and climate change, however, are spread out across the entire planet and arguably may even be disproportionately imposed on less developed parts of the world.

As Upton Sinclair said, "It is difficult to get a man to understand something when his salary depends upon his not understanding it." There are players in the world who may never be in favor of a net-zero state because their livelihoods are currently wrapped up in activities that are antithetical to that goal. Some of the groups that are resistant to changes are relatively obvious. For example, fossil fuel industries have a clear economic reason to be against any substantial movement to renewable and sustainable energy; it will be a move away from their industries. Other groups are less easily recognized but just as opposed to progress. For example, car dealerships are generally not in favor of electric vehicles because EVs require vastly less ongoing maintenance (no oil changes, no transmission fluid checks, no tune-ups, etc.). Repair shop visits, though, are a very large profit center for most dealerships. EVs are seen by many dealerships as a path to lower long-term profits.

By extension, there are governments and regions in the world (e.g., Russia, the United States, and the Middle East) that are either heavily invested or reliant on the revenues flowing from their fossil fuel industries. These political forces can be expected to be resistant to renewable energy replacing those industries. In fact, even if those countries and regions are also invested in renewable energy (in theory, replacing one revenue source with another), it is quite possible that the legislative processes in these locations will tend to be "captured" by the existing fossil fuel industries to a much greater extent than prospective industries.

Recognizing this can help to understand the resistance by these groups not as irrational or delusional but rather the views and positions that are most beneficial to them. It can even make the potential obstacles and opportunities clearer for moving these hostile positions closer to a net-zero world.

10.4 The Paris Agreement on Climate Change

Finally, the relatively neutral position—divorced from economic and political considerations as much as possible—has been represented and periodically updated by the United Nations. The development and approval of the Paris Agreement on Climate Change by the UNFCCC in 2015 and the ratification in 2016 by member countries were major steps forward in the efforts to address climate change. The agreement includes voluntary actions by countries to develop and report plans for nationally determined contributions every two years.

Thus, every two years, this biannual conference of the parties (COP) occurs, with representatives from the participating countries and others meeting and working to address further issues related to climate change. Important decisions are often made at these COP meetings. For example, at COP26, there was an agreement to increase efforts to reduce methane emissions with leadership provided by the European Union and the United States. Over 100 nations have taken the global methane pledge and are taking action to reduce methane emissions by 30% by 2030 (IEA, 2024). At COP28, there was agreement on the need to increase installed renewable energy such that renewable electricity production can be three times greater by 2030 than in 2023 (UNFCCC, 2023). Total pledges to the Green Climate Fund have reached $12.8 billion, with 31 countries making pledges.

Alongside the UN efforts on a global scale, there are many multinational corporations that are attempting to reduce greenhouse gas emissions. Microsoft, for instance, has a goal to reach net zero and even remove carbon from the atmosphere by 2030 (Maslin et al., 2023). Many global companies have pledged to reach net zero by 2050. As of 2023, 88% of greenhouse gas emissions are included in pledges to reach net zero by 2050 (Maslin et al., 2023).

Progress has certainly been made, but not enough (or fast enough). Because of the slow progress in reducing greenhouse gas emissions, there is a need to reach a 50% reduction in greenhouse gas emissions by 2030 to maintain good progress toward the goal of staying below a 1.5 degrees Celsius global temperature increase (Black et al., 2023). The present plan of Nationally Determined Contributions will lead to a reduction of about 11%. Significant changes are needed to reach the temperature goals of the Paris Agreement.

In Chapter 1, we introduced the first law of holes: *If you find yourself in a hole, stop digging.* The UN activities have, over the years, made clear the depth of our climatological hole and how fast we have been digging it. Following the first law of holes, the immediate thing that must be done is to stop digging. Eventually, though, we will need to address the second law of holes: *when you stop digging, you are still in a hole.* A globally warmed planet does not immediately revert to its unaffected state. We will need to fill the hole back in.

On a more hopeful note, we will have help filling our ecological hole back in. The planet's ecosystem (as long as we do not completely disrupt it) will help to restore a changed climate back into balance. There are also manifold positive effects as we reach net zero and go into restoration of global climate quality. Let's look forward to some of these multiplier effects that can happen as we move towards net zero. For example, as Erickson and colleagues (2016) point out, the addition of solar panels to locations such as parking lots can provide energy for charging electric vehicles, reduce demand on the electrical grid and possible use of fossil fuels for energy (the energy for charging is produced on-site, with basically zero transmission loss), and can provide shade and protection for vehicles. Similarly, Erickson and Brase (2020) discuss how lowering greenhouse gas emissions is not only crucial for addressing climate change but will also lead to better air quality and thereby result in very large improvements in global health.

References

Alanazi, F.2023. Electric vehicles: Benefits, challenges, and potential solutions for widespread adaptation. Applied Sciences 13: 6016.

Biden, J. 2021. Federal Sustainability Plan. The White House, Washington, DC. ustainability.gov

Black, S., Parry, I. and Zhunussova, K. 2023. Is the Paris Agreement Working? IMF Staff Climate Note 2023/002, International Monetary Fund, Washington, DC.

Borlaug, B., Salisbury, S., Gerdes, M. and Muratori, M. 2020. Levelized cost of charging electric vehicles in the United States. Joule 4: 1470–1485.

DOE. 2024. Decarbonizing the U.S. Economy by 2050: A National Blueprint for the Buildings Sector. U.S. Department of Energy. energy.gov/eere

Erickson, L.E. 2024. Electric vehicles for environmental sustainability. In Decarbonization Strategies and Drivers to Achieve Carbon Neutrality for Sustainability. Elsevier, Amsterdam, 167–179.

Erickson, L.E. and Brase, G. 2020. Reducing Greenhouse Gas Emissions and Improving Air Quality: Two Interrelated Global Challenges. CRC Press, Boca Raton, FL.

Erickson, L.E. and Ma, S. 2021. Solar-powered charging networks for electric vehicles. Energies 14: 966.

Erickson, L.E. and Pidlisnyuk, V., Eds. 2021. Phytotechnologies for Biomass Production: Sustainable Management of Contaminated Sites. CRC Press, Boca Raton, FL.

Erickson, L.E., Robinson, J., Brase, G. and Cutsor, J., Eds. 2016. Solar Powered Infrastructure for Electric Vehicles: A Sustainable Development. CRC Press/Taylor and Francis Group, Boca Raton, FL. https://doi.org/10.1201/9781315370002.

Gelles, D. 2024. Can we engineer our way out of the climate crisis? New York Times, March 31.

Hampel, C. 2023. CATL presents battery swapping system for trucks. Electrive, June 13. electrive.com.

H.R. 5376. 2022. Inflation Reduction Act of 2022. Congress.gov. Library of Congress. congress.gov/bill/117th-congress/house-bill/5376

Hughes, S. and Zoelle, A. 2022. Cost of capturing carbon dioxide from industrial sources. National Energy Technology Laboratory, Pittsburgh, PA.

IEA. 2024. The global methane pledge. International Energy Agency. iea.org/the-global-methane-pledge/

IRENA. 2024. Record growth in renewables, but progress needs to be equitable. Press Release by International Renewable Energy Agency, March 27. irena.org

Kim, S., Nitzsche, M.P. and Rufer, S.B. 2023. Asymmetric chloride-mediated electrochemical process for CO2 removal from oceanwater. Energy and Environmental Science 16: 2030–2044.

Lanz, L., Noll, B., Schmidt, T.S. and Steffen, B. 2022. Comparing the levelized cost of electric vehicle charging options in Europe. Nature Communications 13: 5277.

Maslin, M.A., Lang, J. and Harvey, F. 2023. A short history of the successes and failures of the international climate change negotiations. UCL Open Environment 5: 8. https://doi.org/10.14324/111.444/ucloe.000059.

McKerracher, C. 2024. Electric vehicle market looks headed for 22% growth this year. Hyperdrive Bloomberg News, January 9. bloomberg.com/news/

Newman, R. and Noy, I. 2023. The global costs of extreme weather that are attributable to climate change. Nature Communications 14: 6103.

Pilorge, H., McQueen, N., Maynard, D., et al. 2020. Cost analysis of carbon capture and sequestration of process emissions from the U.S. industrial sector. Environmental Science and Technology 54: 7524–7532.

Qiang, H., Hu, Y., Tang, W. and Zhang, X. 2023. Research on optimization strategy of battery swapping for electric taxis. Energies 16: 2296.

Rennert, K., Errickson, K., Prest, B.C., et al. 2022. Comprehensive evidence implies a higher social cost of carbon dioxide. Nature 610: 687–692.

Revankar, S.R. and Kalkhambkar, V.N. 2021. Grid integration of battery swapping station: A review. Journal of Energy Storage 41: 102937.

Service, R.F.2024. Firms aim to capture carbon in the oceans. Science 383: 1400–1401.

Sievert, K., Cameron, L. and Carter, A. 2023. Why the cost of carbon capture and storage remains persistently high. International Institute for Sustainable Development, Winnipeg, Manitoba, Canada. iisd.org/

Smith, A.B. 2024. 2023: A Historic Year of U.S. Billion-Dollar Weather and Climate Disasters. NOAA National Centers for Environmental Information. www.ncei.noaa.gov/access/billions/

Tol, R.S.J.2024. A meta-analysis of the total economic impact of climate change. Energy Policy 185: 113922.

UNFCCC. 2015. Paris Agreement. United Nations Framework Convention on Climate Change. United Nations.

UNFCCC. 2023. COP28 Agreement Signals the "Beginning of the End" of the Fossil Fuel Era. United Nations Framework Convention on Climate Change, December 13. unfccc.int/cop28-agreement/

Woody, M., Adderly, S.A., Bohra, R. and Keoleian, G.A. 2024. Electric and gasoline vehicle total cost of ownership across US cities. Journal of Industrial Ecology. https://doi.org/10.1111/jiec.13463.

Wu, Y., Zhuge, S., Han, G. and Xie, W. 2022. Economies of battery swapping for electric vehicles—Simulation based analysis. Energies 15: 1714.

11

Tables

This chapter includes Tables 11.1–11.8 that have been selected because of their value for the content of the book. All of the tables in the book are in this chapter. An attempt has been made to develop tables that have value related to the important topics in the book.

Table 11.1 is important because it provides information on the values of the concentrations of carbon dioxide and methane over time. The atmosphere is global, and concentrations have been increasing in all parts of the world. While the efforts to reduce emissions are local, regional, and by country, the local emissions affect the global values of carbon dioxide and methane. Table 11.2 reports the global carbon dioxide emissions, while Table 11.4 reports the emissions of some countries.

These three tables enable the reader to understand the state of the effort to reduce greenhouse gas emissions. Because of progress in improving sustainable development in many underdeveloped countries, greenhouse gas emissions are still increasing in some countries, as shown in Tables 11.2 and 11.4.

Tables 11.3, 11.5, and 11.6 report progress toward the transition to net zero greenhouse gas emissions. In Table 11.3, the percent values for EVs are small, and there is a need to increase these values by having many more EV sales. The transition to electric vehicles is going better for electric bicycles and electric cars compared to electric buses and electric trucks. The results in Table 11.5 show that electricity generated by onshore wind is global, with significant capacity in many countries.

Electricity production from both wind and solar has been increasing rapidly because of progress in reducing the cost of production. The reduction in the price of solar panels is shown in Table 11.7. This development is very important as part of the effort to transition to renewable energy with an electrical grid that has most of its power coming from wind, solar, hydro, nuclear, and batteries.

Table 11.8 shows the decrease in prices of batteries, which is very important for the transition to EVs and for battery storage of energy.

Additional developments in batteries are expected as part of the progress to reach net zero by 2050. The quantity of batteries needed continues to grow because the transition to electric vehicles includes bicycles, scooters, cars, trucks, buses, and ships. Battery swapping involves charging batteries to replace depleted ones. Batteries are also charged and stored to provide electricity to power the grid.

DOI: 10.1201/9781003396154-11

There is also a brief description and message following each table.

Table 11.1 shows that the concentrations of carbon dioxide and methane in the atmosphere have been increasing because of greenhouse gas emissions.

The goal of net zero is for these values to stop increasing and reach a constant steady-state value.

Table 11.2 reports the annual global carbon dioxide emissions for 2016–2023. The magnitude of the values is very large. The smaller value in 2020 is because the emissions associated with transportation decreased because of less travel in 2020 when COVID-19 was a concern.

TABLE 11.1

Annual Average Concentration of Carbon Dioxide and Methane in the Atmosphere

Year	Carbon Dioxide Concentration (ppm)	Methane Concentration (ppm)
1985	346	1.65
1990	355	1.71
1995	362	1.75
2000	370	1.77
2005	378	1.77
2010	387	1.80
2015	400	1.83
2020	413	1.87
2021	415	1.89
2022	419	1.90
2023	423	1.92

Data from the Global Monitoring Laboratory, Earth System Research Laboratories, National Oceanic and Atmospheric Administration; gml.noaa.gov

TABLE 11.2

Global Annual Carbon Dioxide Emissions

Year	Global Carbon Dioxide Emissions in Billion Metric Tons
2016	35.46
2017	36.03
2018	36.77
2019	37.04
2020	35.01
2021	36.30
2022	37.15
2023	37.40

Data from Global Carbon Dioxide Emissions from Fossil Fuels. Our World in Data; ourworldindata.org/co2

While there are efforts to reduce annual emissions, they have still been increasing each year in many countries.

Table 11.3 provides data for the USA on the percent of sales of that brand that are electric vehicles. Since Tesla sells only EVs, they are not included in this table. All of the top six brands in this list are EVs from Europe.

Table 11.4 contains emissions data for carbon dioxide from several countries. Emissions have increased with time for China and India, while the European Union, Japan, and the United States have made progress in reducing emissions.

Table 11.5 reports onshore wind power capacity for several countries. Wind power capacity has been increasing rapidly because of the low cost of electricity from wind. There is some offshore wind power capacity; however, it is much less and more expensive. Wind power provides more than 40% of the electricity in some parts of the world.

TABLE 11.3

Electric Vehicle Sales as a Percent of Total Brand Sales in 2023 in the U.S.

Brand	Percent EVs
BMW	12.5
VW	11.5
Mercedes	11.4
Audi	11.0
Volvo	10.8
Porsche	10.0
Hyundai	7.2
Cadillac	6.2
Ford	3.8
Kia	3.8
Chevrolet	3.7

Data from Cox Automotive, A record 1.2 million EVs were sold in the U.S. in 2023, according to Kelley Blue Book, January 9, 2024; coxautoinc.com.

TABLE 11.4

Carbon Dioxide Emissions by Country and Year in Billion Metric Tons

Year	China	European Union	India	Japan	United States
2000	3.5	3.4	0.9	1.2	5.9
2005	6.1	3.5	1.1	1.3	5.8
2010	8.8	3.3	1.7	1.2	5.5
2015	10.3	3.0	2.2	1.2	5.0
2020	11.5	2.6	2.2	1.0	4.4
2023	12.6	2.5	2.8	1.0	4.5

Data from the International Energy Agency report CO2 Emissions in 2023; iea.org

Table 11.6 provides sales data on electric bicycles for 2023, with over 2 million sold in Germany. Electric bicycles are inexpensive and efficient. Many people use them for travel and work in many countries. The infrastructure for electric bicycle transportation is good in many European countries.

Table 11.7 contains solar panel prices to illustrate the decrease in price from 1980 to 2022. Because of progress, electricity from solar radiation has decreased in cost, which is very beneficial for society.

TABLE 11.5

Onshore Wind Power Capacity in 2023 by Country in Megawatts

Country	Capacity
China	404,605
United States	147,979
Germany	61,052
India	44,736
Spain	31,021
Brazil	29,135
France	22,194
Canada	16,989
United Kingdom	15,470
Italy	12,308

Data from Onshore wind energy capacity in 2023 by country in Irena report Renewable Energy Capacity Statistics 2024; data from statista.com also.

TABLE 11.6

Sales Volume of Electric Bicycles in Europe in 2023 by Country

Country	Sales Volume in Thousands of Bicycles
Germany	2120
France	781.36
Netherlands	448.36
Italy	416.5
Belgium	379.24
Poland	322.45
Austria	279.78
Spain	245.47
Switzerland	241.91
Denmark	229.12
United Kingdom	180.83
Sweden	170.5
Czechia	144.41

Data from Sales volume of electric bicycles in Europe in 2023 by country provided by Statista; statista.com

One of the important challenges is to manage the production of electricity so that all of the electricity that is generated is used efficiently and effectively.

Table 11.8 shows that lithium-ion battery prices have decreased from 2013 to 2023. The price of the battery for an electric vehicle has been a very significant concern, and the decrease in price has been very beneficial. Lithium-ion batteries have many uses; the decrease in price is also important for electric bicycles and electric buses.

TABLE 11.7

Solar Panel Prices from 1980 to 2022 in $/W

Year	Price in $/Watt
1980	34.80
1985	16.37
1990	11.49
1995	8.11
2000	6.17
2005	4.39
2010	2.32
2015	0.68
2020	0.34
2022	0.26

Data from the International Renewable Energy Agency (IRENA) and Our World in Data; ourworldindata.org

TABLE 11.8

Average Global Price of Lithium-Ion Batteries for Electric Vehicles in U.S. Dollars/Kilowatt-Hour

Year	Price
2013	780
2014	692
2015	448
2016	345
2017	258
2018	211
2019	183
2020	160
2021	150
2022	161
2023	139

Data from Bloomberg NEF as published by Spector, J. and McCarthy, D. 2023. Chart: Lithium-ion battery prices are falling again. Canary Media, December 8, 2023; canarymedia.com.

12

Humans in the Loop: Decision-Making Processes

12.1 Introduction

A million important decisions, big and small, will need to be made along the pathway to net zero. Decisions are made by individuals, families, elected officials, executive officers, organizations, cities, counties, states, provinces, countries, businesses, corporations, universities, public schools, media outlets, United Nations, sports teams, medical offices, and others. There are building codes, taxes, regulations, incentives, agreements, zoning, and other requirements that flow from (or lead to) some of these decisions. When good decisions are made, new products are developed, efficiency is improved, good ideas are communicated, and there is progress toward goals that make our world more livable.

This means that you and others reading this will participate in some of the processes and decisions along this pathway. Indeed, many of you already have made decisions based on the idea of eventually reaching net zero at some future date. The individual commitments and efforts by groups to reduce greenhouse gas emissions are important because they affect progress. There are also efforts to make good decisions by those who are part of the UNFCCC, who attend the annual COP meetings and who work more broadly to make progress toward the goals of the Paris Agreement.

Chapter 10 (economics and policy) included a model of decision-making that is often an initial assumption for how humans make decisions. Specifically, this model compares the costs and benefits of different courses of action and then decides to take the course of action for which there is the greatest net benefit (total benefits minus total costs). This is an expected utility model (or subjective expected utility if the costs and benefits are subject to personal preferences), and it is often the first approximation for evaluating decisions.

We know, however, that human decision-making is much more complex than expected utility maximization. Although some would argue that people are thus inferior decision-makers because of this discrepancy, this is not at all a clear conclusion. Humans are absolutely sensitive to the potential costs and benefits of different outcomes, and this shapes many decisions. But our decisions often also seek to maximize things other than utility, things such as time,

DOI: 10.1201/9781003396154-12

ease, social status, or social conformity. And, of course, people differ in which of these things take priority (including sometimes even the same person, but at different times and in different situations).

The following sections review some of the factors that influence people's decisions about topics that are relevant to the goal of reaching net zero emissions. Sometimes, these factors are related to straightforward cost/benefit economics; sometimes, they are about trying to stand out as an individual (as an early adopter, an adventurous, high-status person); and sometimes, decisions about trying to fit in with everyone else (making the normal, accepted choice). Rather than think of these as contradictory approaches, it can be instructive to compare them to a tradeoff common throughout the natural world: deciding whether to explore the world for new resources or to exploit known resources (Mehlhorn et al., 2015). Exploration can reveal new, untapped opportunities, but it is risky. Exploitation of a known situation is safer, but it risks missing out on better options that were unexplored. Both exploring and exploiting thus carry with them potential benefits and risks, and they are often mutually exclusive options, but they are both important approaches to be able to take.

12.2 Economic Incentives

As discussed in Chapter 10, a straightforward way to nudge people toward particular decision options is to change the costs and benefits somehow. Decreasing the cost of an option or increasing the benefits of an option makes it more attractive; increasing the cost of an option or decreasing its benefit makes it less attractive. Some of these cost/benefit changes occur organically through things like supply and demand changes or technological innovation. Often, though, these changes to the incentive structure of decision options are done intentionally by adopting various policies.

There have been numerous successes in changing decision patterns overall through providing incentives. Tax incentives include both levying additional taxes on less desirable things, such as a carbon tax, and tax deductions, such as a rebate for purchasing an EV. In general, people have a lot of experience with incentives, and they work across myriad contexts and countries. Pro-environmental incentives range from return deposits of a few cents on bottles and aluminum cans to systemic incentive structures designed to aggressively reduce emissions, such as in the case of Norway (Erickson and Brase, 2020).

There are local, state, and federal incentives to install wind and solar systems for electricity generation in many locations. As people take advantage of these incentives, they add renewable energy to the grid and reduce emissions in those locations. There are incentives (tax breaks, discounts, etc.) in many locations for fitting buildings with more efficient thermostats, better insulation, and using less water. At a population level, these tend to be effective, and when carefully implemented, they have the desired effect of reducing emissions on broader scales.

At the individual level, everyone makes decisions daily for a number of different purposes. For example, in August of 2013, I (LEE) purchased a Toyota plug-in hybrid Prius because I wanted the experience associated with driving powered by electricity. This Prius had an electric range of about 13 miles (20 km), which was very appropriate for my situation since my commute to work was only about 1 mile, and many days, the total distance that I drove was less than 12 miles. Because it is a hybrid, the gasoline engine comes on when it is needed. I selected this car because there was a local dealership for service, it could serve many of my transportation needs, and the cost was reasonable for the benefits that I wanted. An incentive that helped was that I also received a deduction on my federal income taxes. The car has many features that I like. I use a standard 120 volt, 30 amp circuit in my garage to charge the electric battery, which powers the electric motor and is about 4 kWh. As of May 2024, the car is more than 10 years old and has been driven almost 100,000 miles. I have been very satisfied with my experiences as an owner of this car.

Changing the incentive structures can work not just for individuals but also for companies and other organizations. Companies in the energy sectors can get research and development support, subsidies, loan guarantees, and other forms of incentives. The good news is that the U.S. Energy Information Administration (2023) found that federal support for renewable energy of all types more than doubled, from $7.4 billion in the 2016 fiscal year to $15.6 billion in 2022. On the other hand, US subsidies for fossil fuels reached $7 trillion in 2022, according to the International Monetary Fund (Black et al., 2023). The fact that the US federal government is promoting both fossil fuels and renewable energy at the same time illustrates a couple of important points. First, the federal government is not a monolith; different factions within the government can be working to incentivize different decisions and behaviors—even somewhat conflicting ones. Second, intentional changes to incentive structures can sometimes run into problems: either there are countervailing incentives that make them ineffective, or the new incentives turn out to reward behaviors other than what were the intended outcomes (a phenomenon sometimes called "perverse incentives").

Here are a couple of examples of incentives that turned out to be misaligned and sometimes even perverse incentives:

- While automobile manufacturers have been steadily ramping up the production of EVs, dealerships have been remarkably slow to promote consumers purchasing them. Many dealerships lacked qualified or knowledgeable salespeople to sell EVs, and in other cases, they might even discourage potential customers from purchasing an EV. Why would this happen? To understand this, it is important to realize that servicing and part sales make up a large part of dealership revenues. EVs, which have fewer moving parts and therefore require less servicing, are projected to generate only 60% of the aftermarket business that gas-powered vehicles provide (Fischer et al., 2021). The prospect of lost future revenue creates a perverse incentive for dealerships to continue selling gas-powered vehicles rather than EVs.

- There have been vehicle efficiency standards in the United States for nearly 50 years, called corporate average fuel economy (CAFÉ) standards. These involve penalties to automobile manufacturers so that producing inefficient vehicles is more expensive. Over the years there have been many changes to these CAFÉ standards, both to address new technologies and to better fit with the goals of successive government administrations. An amendment to the CAFÉ standards in 2011 changed the fuel economy targets to be based on the footprint (wheelbase by track width) of vehicles, with lower fuel-economy targets for larger vehicles. This, however, created a perverse incentive for automakers to increase the size of some vehicles—notably trucks—to qualify for lower fuel economy targets rather than make them more efficient (Whitefoot and Skerlos, 2012).

Workman et al. (2023) provide a good case study on general decision-making for net zero emissions, including guidelines on how to make good decisions and then put them into practice. Notably, Workman and colleagues included many different methods to identify how to implement constructive and effective decision-making settings. They conducted a literature review and interviews and held policy workshops to address gaps associated with net zero decision support in developing climate policy. The results include a set of three recommendations:

1. Enhance collaboration among decision-makers, those who make policy, and other stakeholders;
2. Identify the gaps related to the decision-making and address the gaps;
3. Co-create effective processes to include decision-support tools in policy development using a participatory approach with the inclusion of diverse viewpoints and values.

12.3 Normalizing Net Zero

Other than the incentives of costs and benefits, how do people decide what to do? This turns out to be a more significant question than it might appear at first. There are many, many situations in real life where the full costs and benefits—both immediately and into the future—are unclear. We often lack the information we need to make decisions based on these factors, either because our knowledge is limited or because the information simply does not exist. So what then?

A very common solution to making decisions under such uncertainty is to continue doing whatever has worked before, either for us or for other people around us. This corresponds to the strategy of "exploit" (rather than "explore") that was mentioned earlier. People are comfortable continuing to do what they have done before (assuming it still works), and they are often willing to follow the examples of others as a guide to other decisions and behaviors that work.

This tendency to conform to norms—doing what others are doing because that appears to be the normal thing to do—can occur organically, or it can be leveraged to help gradually move people to more environmentally friendly behaviors. The key is to make the environmentally friendly behaviors of others highly visible so that others see them as normative options for themselves.

Understanding this aspect of human decision-making helps to understand a number of phenomena. For example, one study of what factors people used when thinking about whether to buy an EV was their perceptions of how prevalent EVs were in general already (Brase, 2018). Other people who have EVs already increases the chances of the next person also buying an EV. Another study (Graziano and Gillingham, 2015) found that the likelihood of people installing solar panels on their home was significantly influenced by the presence of solar panels on the roofs of their neighbors. Other people having solar panels increased the chances of a neighbor also putting up solar panels. Lastly, there have been a number of studies on the use of information about the energy use of one's neighbors, compared to your own usage (e.g., Clayton, 2009). This feedback information can be on monthly bills (e.g., smiley faces for using less energy than your neighbors) or even on a smart thermostat. Generally, energy consumption decreases in households that receive this type of comparison information.

Infrastructure and building projects are significant sources of greenhouse gas emissions. Much effort is needed to reduce emissions associated with buildings, including homes. Fang and colleagues (2023) review five decision-making stages that are needed for transitioning buildings to have net zero emissions. The first stage is to define the scope, goals, and boundaries of the project. The second stage is to define the methodology and the methods to use to reduce emissions. The third stage is to review the data and choices to reduce emissions. The fourth stage is the analysis of the alternatives. The fifth stage is to create a visualization product to communicate the alternatives being considered and their features, including the estimated emissions associated with the alternatives (Fang et al., 2023).

The U.S. National Academies has produced a recent report, "Greenhouse Gas Emissions Information for Decision Making," which is intended to provide guidance on the use of greenhouse gas information for decision-making (NASEM, 2022). The report provides a framework to evaluate emissions information and inventories as well as to understand the transparency and continuity of reported data. A common evaluation framework is beneficial for the analysis of emissions data. Usability, information transparency, evaluation, validation, completeness, inclusivity, and communication are important concerns in the report and evaluation criteria. Project data accessibility is also included in this evaluation framework.

12.4 Standing Out From the Crowd

There are times when some people do not use their past behavior or the behaviors of others as their decision-making guides. Instead, they strike out on a path less taken and "explore" new options (rather than "exploit" their current situation). There may, after all, be options that are better than the one everyone else has taken, and how would you know an alternative is better if you don't try it?

Individuals who are "early adopters" of technologies are people who are exploring new opportunities. Whether it is home solar panels, EVs, or smart thermostats, early adopters are trying out new possibilities because they see something potentially more valuable in that new space. It may be because their calculations of costs and benefits are different; they perceive larger potential benefits and fewer possible costs. Or, it could be that they are looking toward some other goal such as time savings, ease of use, or social status.

For example, Griskevicius and colleagues (2010) showed that people were more likely to engage in pro-environmental behaviors when they were set up in such a way that it increased their social status and reputation. They called this "conspicuous conservation," which is a turn of phrase based on an older phenomenon known as conspicuous consumption (when a person with a lot of resources expends them in a very public way in order to demonstrate the extent of their wealth; see Griskevicius et al., 2007). Crucially, this technique worked only when the environmentally responsible behavior was made visible to others; it did not work when the behavior was private.

Similar motivations may explain phenomena such as the popularity of Tesla cars (particularly when they were difficult to obtain) and early adopters of home solar panels. Conversely, it can explain why many other environmentally responsible behaviors (conserving electricity and water, recycling, etc.) are less popular; they are often not visible and are rarely based on demonstrating the ability to achieve something that others cannot do.

Just like with changes to incentive structures, though, setting up these situations can be tricky. They can even backfire, at least for certain circles of people. Contrarians can emerge when new, pro-environmental technologies are seen as progressive status symbols. Witness the phenomenon of "rolling coal" that emerged after the initial popularity of hybrid and EV cars. Some individuals deliberately altered their large vehicles to spew out large amounts of exhaust as a reaction against less polluting cars. This likely also was viewed as a public display of status and reputation, albeit for a quite different target audience.

The technique of showing people how their own energy consumption compares to the usage levels of their neighbors (discussed previously) also has limits. Further work (Dora et al., 2013) found that this type of feedback is two to four times more effective in households that identify as politically liberal than in households that identify as conservative. Arguably, this contrarian position may have also been behind the reticence of the US Postal Service Postmaster General, Louis DeJoy, (appointed by a conservative president) to replace the majority of the service's aging fleet with electric vehicles (Stark, 2022). DeJoy instead wanted to acquire more gas-powered vehicles until intense pressure from the subsequent administration insisted on more EVs for the USPS fleet (USPS, 2022).

12.5 Collective Behavior for a Common Resource

Our individual decisions are made based on personal evaluations of costs and benefits, both in terms of economics and also in terms of things like social status, saving time, and fitting in with the people around us. Cumulatively, those individual decisions add up to much of our overall response to climate change. One outcome is that we can collectively take action to address air, water, and land pollution. Or we can collectively do nothing as a society, shrug our shoulders, and turn our backs on the problem of global climate change. We are currently somewhere in between these extremes, with a key question being how we move toward the former outcome rather than the latter.

In other words, how do we transition from early adopters to accepted behaviors to normative behaviors and finally to collective behaviors? We are still in the "early adopter" stage in some areas, such as agricultural transitions and smart electrical grids. We appear to be somewhere in the second step of this sequence in some other ways, such as EV adoption; early adopters have been giving way to new car options, increasingly becoming EVs. A critical idea after that is the move to normative behavior when the normal default is the more environmentally responsible decision.

When we reach the point at which a new behavior is perceived to be the normative one, there is often a "tipping point." After this tipping point, the proportion of people adopting the new behavior increases rapidly because it is seen as the new default (Gladwell, 2000; Frank, 2020).

A recent report (Systemiq, 2023) provides a set of guidelines for how to move many different sectors towards a net zero emissions state by pushing them up to and through their tipping points.

It is important to get everyone on board for a net zero emissions future because collective action is required, not just because it is an "all hands on deck" situation that requires a huge amount of effort (although that is true). Collection action is also important because the underlying situation being addressed is one of several interrelated renewable common pool resources: our planet's air, soil, and water systems. We all share these things by virtue of living on Earth, and people who overuse, abuse, or do not do their fair share to maintain these common resources must be adamantly identified as abusers of these public goods. Chapter 15 of our earlier book addresses the commons issue that is associated with greenhouse gases in the atmosphere (Erickson and Brase, 2020).

The Earth is the only planet available for implementation of the Paris Agreement on climate change. Greenhouse gas emissions impact the entire Earth. Each person has a global impact, and each country has a significant impact. The emissions associated with war have global impacts. The nationally determined contributions plan of UNFCCC allows each country to develop its own plan to reduce emissions. Past efforts to address commons problems have used this approach and had success (Erickson and Brase, 2020).

For all the previous reasons, education is key. Everyone needs to understand what is at stake in terms of polluting emissions, global climate change, and net zero emission status as an essential step to address these problems. Everyone needs to understand the global contract that is entailed by global common resources. And, fortunately, education related to pathways to net zero greenhouse gas emissions has been shown to be beneficial.

The science and technology behind both climate change and net zero are natural topics for STEM education in schools. Additionally, the actions that can be taken to reach net zero need to be included in educating the public in all countries. Information on the transition to sustainable energy, electric vehicles, and green hydrogen needs to be included in public education. Children in public schools, university students, adults, community leaders, elected representatives, and administrators all need to have an understanding of the many aspects associated with sustainable development, what is needed for the transition to net zero emissions, and the pathways that are available to reach net zero. Each person needs to have opportunities to learn these topics.

The information is already out there (in addition to this book). The annual COP meetings of the UNFCCC provide public education through the news media as well as direct information for all who attend. Reports are prepared, distributed, and posted on the online. The US National Academies for Sciences, Engineering, and Medicine have a series of consensus documents and books, generally available for free online (www.nationalacademies.org/our-work), that includes several resources on how to successfully and rapidly decarbonize the United States (i.e., reduce greenhouse gas emissions). The latest report from this group (National Academies of Sciences, Engineering and Medicine, 2024) includes about 80 specific recommendations for specific immediate steps. Of all the reasons people may give for failing to make environmentally responsible decisions, not knowing better should never be a realistic option.

References

Black, S., Parry, I. and Vernon, N. 2023. Fossil Fuel Subsidies Surged to Record $7 Trillion, August 24. www.imf.org/en/Blogs/Articles/2023/08/24/fossil-fuel-subsidies-surged-to-record-7-trillion

Brase, G.L. 2019. What would it take to get you into an electric car? Consumer perceptions and decision making about electric vehicles. The Journal of Psychology 153(2): 214–236. https://doi.org/10.1080/00223980.2018.1511515.

Clayton, M. 2009. Energy use falls when neighbors compete. The Christian Science Monitor, September 30. www.csmonitor.com/Technology/2009/0930/energy-use-falls-when-neighbors-compete

Costa, D.L. and Kahn, M.E. 2013. Energy conservation "Nudges" and environmentalist ideology: Evidence from a randomized residential electricity field experiment. Journal of the European Economic Association 11(3): 680–702. https://doi.org/10.1111/jeea.12011.

Erickson, L.E. and Brase, G. 2020. Reducing Greenhouse Gas Emissions and Improving Air Quality: Two Interrelated Global Challenges. CRC Press, Boca Raton, FL.

Fang, Z., Yan, J., Lu, Q., et al. 2023. A systematic literature review of carbon footprint decision-making approaches for infrastructure and building projects. Applied Energy 335: 120768.

Fischer, M., Kramer, N., Maurer, I. and Mickelson, R. 2021. A turning point for US auto dealers: The unstoppable electric car, September 23. www.mckinsey.com/industries/automotive-and-assembly/our-insights/a-turning-point-for-us-auto-dealers-the-unstoppable-electric-car

Frank, R.H. 2020. Under the Influence: Putting Peer Pressure to Work. Princeton University Press, Princeton, NJ.

Gladwell, M. 2000. The Tipping Point: How Little Things Can Make a Big Difference. Little Brown, New York, NY.

Graziano, M. and Gillingham, K. 2015. Spatial patterns of solar photovoltaic system adoption: The influence of neighbors and the built environment. Journal of Economic Geography 15(4): 815–839. https://doi.org/10.1093/jeg/lbu036.

Griskevicius, V. and Tybur, J. 2010. Going green to be seen: Status, reputation, and conspicuous conservation. Journal of Personality and Social Psychology 98(3): 392–404.

Griskevicius, V., Tybur, J.M., Sundie, J.M., Cialdini, R.B., Miller, G.F. and Kenrick, D.T. 2007. Blatant benevolence and conspicuous consumption: When romantic motives elicit strategic costly signals. Journal of Personality and Social Psychology 93(1): 85–102. https://doi.org/10.1037/0022-3514.93.1.85.

Mehlhorn, K., Newell, B.R., Todd, P.M., Lee, M.D., Morgan, K., Braithwaite, V.A., Hausmann, D., Fiedler, K. and Gonzalez, C. 2015. Unpacking the exploration—exploitation tradeoff: A synthesis of human and animal literatures. Decision 2(3): 191–215. https://doi.org/10.1037/dec0000033.

NASEM. 2022. Greenhouse Gas Emissions Information for Decision Making: A Framework Going Forward. National Academies Press. https://doi.org/10.17226/26641.

National Academies of Sciences, Engineering, and Medicine. 2024. Accelerating Decarbonization in the United States: Technology, Policy, and Societal Dimensions. The National Academies Press, Washington, DC. https://doi.org/10.17226/25931.

Stark, L. 2022. DeJoy defends plan to replace USPS fleet with gas-powered trucks, citing "dire financial condition". CNN, February 8. https://edition.cnn.com/2022/02/08/politics/dejoy-defends-usps-gas-vehicles-electric-climate/index.html

Systemiq. 2023. The Breakthrough Effect: How to Trigger a Cascade of Tipping Points to Accelerate the Net Zero Transition, January 2023. www.systemiq.earth/wp-content/uploads/2023/01/The-Breakthrough-Effect.pdf

US Energy Information Administration. 2023. Federal Financial Interventions and Subsidies in Energy in Fiscal Years 2016–2022, August 1. www.eia.gov/analysis/requests/subsidy/

USPS. 2022. USPS Intends to Deploy Over 66,000 Electric Vehicles by 2028, Making One of the Largest Electric Vehicle Fleets in the Nation, December 20. https://about.usps.com/newsroom/national-releases/2022/1220-usps-intends-to-deploy-over-66000-electric-vehicles-by-2028.htm

Whitefoot, K.S. and Skerlos, S.J. 2012. Design incentives to increase vehicle size created from the U.S. footprint-based fuel economy standards. Energy Policy 41: 402–411.

Workman, M., Heap, R., Mackie, E. and Connon, I. 2023. Decision making for net zero policy design and climate action: Considerations for improving translation at the research policy interface: A UK Carbon Dioxide Removal case study. Frontiers in Climate 5: 1288001.

13

International Sustainable Development: Collective Action, War, and Peace

13.1 Introduction

Net zero emissions is a global target: to get to where the total emissions produced across the planet are no more than those emissions being removed globally. There are, of course, a multitude of more local actions at the levels of countries, states, cities, and even individuals, and these all contribute to this global goal. But in the end, the net zero goal is in service to particular ends—stopping further pollution and global climate change—that are planet-wide common good situations. When the overall global planet warms up, when air is polluted, and when oceans rise from melting ice everyone experiences these effects of global common goods having deteriorated.

So many of the things we have covered in this book—renewable energy sources, more efficient and reliable energy grids, the electrification of transportation, homes, and agricultural processes—are, in this respect, all simple means to an end. We need to stop overusing the common pool resource that is our planetary ecosystem.

Because there are chronically selfish incentives to overuse resources for individual gain, the responsible use of common pool resources usually needs some type of collective action. In simple terms, this means rules, what actions are expected of everyone, and what the consequences are for people who break those rules. (Depending on the nature of the common resource, these rule breakers can be referred to as defectors, free riders, or simply cheaters.) Fortunately, it seems that there is remarkable consensus in general on how people *should* behave. Most people faithfully follow the guides for collective action, and even most of the people who do not do so are aware nevertheless that they are supposed to.

There are many efforts to work cooperatively to reduce global greenhouse gas emissions and make progress toward specific sustainable development goals. Two prominent efforts among these are the UN Climate Change Conference of the Parties (COP) and the Earth Charter. The United Nations provides many opportunities to work cooperatively to advance progress toward net zero greenhouse gas emissions through the annual United Nations COP and other

DOI: 10.1201/9781003396154-13

activities related to the goals of the Paris Agreement on Climate Change. The Earth Charter (https://earthcharter.org/) grew out of calls by the UN over 35 years ago for a set of fundamental sustainable development principles, and a commission formed to answer that call. The Earth Charter includes 16 principles for the transition to sustainable living and human development, organized under four central themes.

Another key aspect of international sustainable development is that it must involve everyone. A huge amount of attention is paid, including in this book, to the emissions and quality of life in WEIRD (Western, Educated, Industrialized, Rich, & Democratic) places. These are, to be sure, the locations where a majority of emissions have historically come from, are currently coming from, and where the cutting-edge research on addressing climate change is happening. But the rest of the world, which is most of the world, is rapidly developing and getting more like WEIRD countries every day.

There is a need to foster development and improve the quality of life in many countries while also making progress toward sustainable development goals and reducing greenhouse gas emissions. In the meantime, there are many people who are attempting to find a better place to live because of the poor quality of life where they currently reside. If we fail to support improvements in countries worldwide, or if we fail to address global climate change, climate/economic migration will only become more common. Efforts to improve the quality of government and quality of living through education, application of Earth Charter principles, and community programs to address poverty and other human needs have the potential to be beneficial in countries worldwide. The United Nations programs, government programs, and programs of non-government organizations (NGOs) are also helpful in these efforts.

In this chapter, the importance of including physical, environmental, social, financial, and political issues in the efforts to reach net zero greenhouse gas emissions is considered. Making progress on the sustainable development goals is a closely related challenge. All countries should work cooperatively and with urgency to improve the quality of life for their citizens and to reach net zero for future generations of citizens.

13.2 Sustainable Development

The United Nations established a set of 17 Sustainable Development Goals (SDGs) that constitute one of the most commonly referenced benchmarks to evaluate how nations, regions, and the world overall are doing. Part of the appeal of these SDGs is that they are operationally defined; there are measurable outcomes that define both what the current situations are and how much things have improved (or not). Overall, progress has been made. One of the recent publications related to these SDGs is the Special Edition of the Sustainable Development Goals Report 2023 (United Nations, 2023). Progress related to these goals, though, has been impacted recently by the COVID-19 pandemic and the Russian invasion of Ukraine. Specifically, funding for sustainable development goals has been reduced because of financial diversions to address COVID-19, migrations of populations, and wars. The UN General Assembly met in September 2023 for an SDG summit and reaffirmed the commitment to effectively implement the SDG agenda and uphold the principles in it (SDG Summit, 2023).

The financial support for the SDGs includes the Political Declaration that 0.7% of the gross national income of developed countries be contributed to sustainable development as part of the SDG Official Development Assistance (SDG Summit, 2023). The amount of financial support needed for SDGs is about $500 billion/year. A global carbon tax to provide funding for the SDGs and net zero programs would be very beneficial.

World leaders and their representatives reaffirmed their commitment to the 2030 Agenda for Sustainable Development at the SDG summit, including planning for transformative actions through science, technology, and innovation to address poverty, improve food security, increase access to health, and enhance the quality of education. Proposed future events include 2 UN peace conferences (on Ukraine and Palestine) and a conference on financing for sustainable development (SDG Summit, 2023). September 2023 was considered to be the halfway point to the 2030 date for achieving the SDGs.

There is a section at the end of this chapter specifically on the topic of peace—as opposed to war—and issues of sustainable development. For now, it is sufficient to note that peaceful coexistence is included as part of sustainable development because of the importance of effective institutions, equality, justice, and stability; all things that are casualties of war. The UN Sustainable Development Report provides 17 proposed goals and targets. Progress on each of these principles is based on the 2023 United Nations Convention on Climate Change Report

- <u>SDG 1 is to end poverty.</u> There was progress in reducing poverty from 2015 to 2019, but COVID-19 had a negative impact on poverty starting in 2020. Global extreme poverty was 8.8% in 2021.
- <u>SDG 2 is to end hunger and improve food security.</u> In 2022, 9.2% of the population was experiencing chronic hunger, and 29.6% had moderate to severe food insecurity, with an increase from 2019 because of COVID-19. The war in Ukraine has also resulted in increased food prices. The situation was better in 2019.
- <u>SDG 3 is to improve health and well-being.</u> There has been some progress related to the efforts to improve health in many countries; however, childhood vaccinations have declined. Assisted childbirth by skilled helpers has increased.

- SDG 4 is on quality education. COVID-19 had a major impact on education, and there is a need for more funding for education in many countries. Worldwide primary school completion has increased to 87%, and upper secondary completion to 58% as of 2021.
- SDG 5 is on gender equality. The progress is poor, with only 15.4% of indicators on track. There are many systemic barriers that need to be addressed through policy changes.
- SDG 6 is on clean water and sanitation. There has been significant progress with access to safe drinking water; as of 2022, 73% have safely managed drinking water services, 57% have safely managed sanitation services, and 75% have basic hygiene services.
- SDG 7 is on affordable and clean energy. As of 2021, 91% of people have access to electricity; 30% of electricity is from renewable sources; 29% of people are still using inefficient and polluting cooking systems. There is progress, but the rate needs to increase.
- SDG 8 is on decent work and economic growth. Global unemployment has been declining, and economic growth has been increasing as of 2021; 76% of adults had bank accounts; youth unemployment is a problem that needs to be addressed.
- SDG 9 is on industry, innovation, and infrastructure. There has been progress in global manufacturing growth; the per capita global manufacturing value was $1879 in 2022. In 2022, 95% of people were within reach of a mobile broadband network.
- SDG 10 is on reducing inequality in and among countries. The COVID-19 pandemic caused a large increase in income inequality between countries. From 2009–2022, many countries reported greater income growth for the poorest 40% of the population. In 2022, there were 34.6 million refugees; this is the largest number ever documented.
- SDG 11 is on sustainable cities and communities. In 2022, 51.6% of the global urban population had convenient access to public transportation. About 1.1 billion people live in slum-like conditions in cities, and 1 billion do not have all-weather roads. About 55% of the global population lives in cities. There is a need to improve basic services, housing, green spaces, and transportation.
- SDG 12 is on responsible consumption and production. Food waste is a global issue that continues to be very important because of the need for food and the magnitude of the problem of 120 kg/person per year. It is an issue in all countries, and all people need to improve their stewardship of food. Education on sustainable management of natural resources is needed.
- SDG 13 is on climate action. The plan for developed countries to provide $100 billion/year to developing countries needs to be continued

and achieved in order for developing countries to reduce emissions. Education on climate change and actions that are needed to reduce greenhouse gas emissions should be improved and expanded in many places.

- SDG 14 is on oceans, seas, and water. There is a continuing need to address coastal eutrophication, plastic pollution, and algal blooms. Illegal fishing continues to be a problem. There has been progress in marine agreements for ocean protection of marine biodiversity.
- SDG 15 is on sustainable use of land. The emphasis on ecosystem restoration and the Kunming-Montreal Global Biodiversity Framework provide pathways for progress for SDG 15. There is a great need to end deforestation and restore degraded land. Funding increases for biodiversity conservation have been beneficial, but ecosystem restoration efforts need to be expanded in many countries.
- SDG 16 is on peace, justice, and strong institutions. This is a very important SDG because the annual cost of violence and war is about $14 trillion/year, and it has pervasive impacts across many of the SDGs. In 2022, about 108 million people were displaced.
- SDG 17 is on partnerships for the goals. Official development assistance increased in 2022; however, this included increased expenditures because of more refugees and spending related to the war in Ukraine. Because of COVID-19, many developing countries have debt problems. In 2022, 5.3 billion people made use of the Internet.

The efforts to make better progress toward improving the quality of life in all countries are continuing. The United Nations SDG summit in 2023 has been beneficial, but there is a need for everyone to work together to address important issues related to advancing sustainable development.

13.3 United Nations COP28

In 2023, the Conference of the Parties No.28 (COP28) was held in the United Arab Emirates from 30 November to 13 December 2023. There is a report on the Conference, the results of the first global stock-taking, and some of the decisions adopted by the Conference (UNFCCC, 2024).

This summary from COP28 indicates that global greenhouse gas emissions exceed those needed to meet the Paris Agreement temperature goal. Global efforts are needed to triple renewable energy capacity by 2030, increase efforts to decrease the use of coal, reduce the use of petroleum fuels for energy, reduce methane emissions, reduce carbon emissions from road transport, and phase out fossil fuel subsidies. One of the positive items in the summary is the significant reductions in costs for electricity from wind and solar sources and battery storage. There is a need to halt and reverse deforestation and improve sustainable management of forests, as well as to increase funding for forest ecosystem improvement (UNFCCC, 2024).

In order to reach net zero emissions by 2050, the estimated global financial investment for clean energy is $4.3 trillion/year up to 2030 and $5.0 trillion/year from 2030 to 2050. An increase in funding support from developed countries to developing countries is needed to fully support clean energy progress in developing countries (UNFCCC, 2024).

Capacity building is important in the effort to reach net zero by 2050, and good progress has been made with the help provided by the Paris Committee on Capacity-building and others. Efforts are to continue to help developing countries prepare and implement their nationally determined contributions through the sharing of knowledge with workshops and other assistance efforts (UNFCCC, 2024).

International cooperation and multilateral efforts in support of advances in sustainable development and international trade are needed for climate action and implementation of climate policies to achieve the long-term goals of the Paris Agreement. Active participation is encouraged by non-party stakeholders from civil society, business, research institutions, financial institutions, cities, and youth (UNFCCC, 2024).

There are plans for dialogues to discuss progress toward the goals, information on good practices, and effective nationally determined contributions, followed by a report. Future plans include a set of activities for "Road Map to Mission 1.5," in which there will be an effort to stimulate ambition and enhance action to keep the goal of limiting the temperature increase of 1.5 degrees C within reach (UNFCCC, 2024).

The preparation and information gathering for the second global stocktaking will begin with activities for COP30 in November 2026 and conclude two years later at COP32 (UNFCCC, 2024). The summary of these plans for upcoming events includes a reminder of the goal to have early warning systems for all by 2027 because of extreme weather and the benefits of taking action to avoid high winds and flood waters. The summary emphasizes the implementation of sustainable agriculture, ecosystem restoration, improved resilience, global solidarity, and the well-being of all people (UNFCCC, 2024). The summary also includes information on a number of decisions that were made at COP28 related to adaptation, the just transition work program, the mitigation ambition and implementation work program, and the operationalization of some new funding arrangements (UNFCCC, 2024). The COP28 reports include one funding mechanism based on nationally determined contributions (UNFCCC, 2023a) and one on long-term low-emission development strategies (UNFCCC, 2023b).

The COP28 report and past reports show that global greenhouse gas emissions have been continuing to increase annually because some countries have been increasing emissions due to ongoing development activities.

The current results from COP28 indicate that the annual growth in greenhouse gas emissions will end between 2023 and 2030. The projection is that emissions in 2030 will be 2% below those in 2019 (UNFCCC, 2023a); however, that decrease needs to be about 43% in order to meet the goals of the Paris Agreement. Finance is an important issue because there is a need for funds to pay for large-scale transitions such as the use of renewable electricity, green hydrogen production, and electric vehicles. One of the projections based on the review of nationally determined contributions and other available information is that global greenhouse gas emissions could be reduced by about 63% in 2050 compared to values in 2019 (UNFCCC, 2023b).

13.4 Earth Charter and Importance of Peace

There are not many people who are unconditional proponents of war. War not only produces a lot of death but also tends to produce a lot of environmental degradation and is economically expensive. It is true that there is something in human nature that lends itself to violent conflict at times, but we as a species need to continue promoting instead what Abraham Lincoln called "the better angels of our nature."

In addition to the economic windfalls and lack of young people dying due to warfare, there are large environmental benefits from reducing the scale, frequency, and wantonness of wars. Finding peaceful alternatives to warfare will aid in the reduction of greenhouse gas emissions, halting climate change, and leave us better able to increase funding for these (and other) important global challenges.

Thus, the Earth Charter (2000) is considerably more far-reaching than UN sustainability goals. The Earth Charter's 16 ethical principles are:

1. Respect Earth and life in all its diversity.
2. Care for the community of life with understanding, compassion, and love.
3. Build democratic societies that are just, participatory, sustainable, and peaceful.
4. Secure Earth's bounty and beauty for present and future generations.
5. Protect and restore the integrity of Earth's ecological systems with special concern for biological diversity and the natural processes that sustain life.
6. Prevent harm as the best method of environmental protection and, when knowledge is limited, apply a precautionary approach.

7. Adopt patterns of production, consumption, and reproduction that safeguard Earth's regenerative capacities, human rights, and community well-being.
8. Advance the study of ecological sustainability and promote the open exchange and wide application of the knowledge acquired.
9. Eradicate poverty as an ethical, social, and environmental imperative.
10. Ensure that economic activities and institutions at all levels promote human development in an equitable and sustainable manner.
11. Affirm gender equality and equity as prerequisites to sustainable development and ensure universal access to education, health care, and economic opportunity.
12. Uphold the right of all, without discrimination, to a natural and social environment supportive of human dignity, bodily health, and spiritual well-being, with special attention to the rights of indigenous peoples and minorities.
13. Strengthen democratic institutions at all levels and provide transparency and accountability in governance, inclusive participation in decision-making, and access to justice.
14. Integrate into formal education and life-long learning the knowledge, values, and skills needed for a sustainable way of life.
15. Treat all living beings with respect and consideration.
16. Promote a culture of tolerance, nonviolence, and peace.

These principles are organized under four subject areas, which are 1. Respect and care for the community of life, 2. Ecological integrity, 3. Social and economic justice, 4. Democracy, nonviolence, and peace. They are more aspirational than the UN goals and less focused on quantitative metrics to evaluate progress toward the desired state. Recent progress toward the Earth Charter principles, net zero greenhouse gas emissions, and sustainable development in general has been challenging. Time, attention, and resources have been stressed by COVID-19, the quality of some current governments, violence in many countries, and wars.

Our expectations and hopes are that these will be transitory, and when we collectively return our time, attention, and resources to sustainable development and net zero emissions, there will be an ever greater pressure from all citizens to take major actions. Our quality of life depends on integrity, personal conduct, following principles such as those in the Earth Charter, the quality of government, regulations, implementation of justice, the educational system, and environmental quality. If every country expected and encouraged everyone to live according to the principles in the Earth Charter, this would be beneficial in the efforts to reach net zero emissions and increase progress to meet the SDGs. And, of course, what countries expect of their populations is, to a large extent, based on what that population expects and demands of that country.

One last note for reference and—possibly—action. The Institute for Economics and Peace (IEP) has developed the Positive Peace Index (PPI) and the Global Peace Index (GPI) (IEP, 2024). Countries with higher PPI scores have good economies, higher well-being, better social cohesion, and greater satisfaction. Training programs by the IEP to improve positive peace are available (IEP, 2024). The estimated annual cost of violence of about $14 trillion/year is from estimated values reported by IEP. There is a need to increase efforts to end war and find peaceful solutions to disagreements through the United Nations, peace organizations, and all other efforts.

References

Earth Charter. 2000. Earth Charter International. Charter for Compassion, Bainbridge Island, Washington. charterforcompassion.org/350-org/earth-charter/

IEP. 2024. Positive Peace Report 2024. Institute for Economics and Peace, Sydney, March. visionofhumanity.org/resources

SDG Summit. 2023. Informal Summary of the United Nations General Assembly SDG Summit on 18–19 September 2023. un.org/en/conferences/sdgsummit2023/

UNFCCC. 2024. Report of the Conference of the Parties Serving as the Meeting of the Parties to the Paris Agreement on Its Fifth Session, held in the United Arab Emirates from 30 November to 13 December 2023.

UNFCCC. 2023a. Nationally Determined Contributions under the Paris Agreement. United Nations Framework Convention on Climate Change Report. FCCC/PA/CMA/2023/12.

UNFCCC. 2023b. Long-Term Low-Emission Development Strategies. United Nations Framework Convention on Climate Change Report. FCCC/PA/CMA/2023/10.

United Nations. 2023. The Sustainable Development Goals Report 2023: Special Edition. United Nations. unstats.un.org/

14

Conclusions: Recent Developments

14.1 Introduction

All the diverse efforts to reduce greenhouse gas emissions and reach net zero are ongoing processes right now. The global transition to renewable energy production is well underway, although it is still ramping up. The transitions of energy use in transportation (electric vehicles, trucks, bikes, and even planes) and housing (heating, cooling, and lighting) are also accelerating as we watch. Transitions in the agricultural sector (net zero farming and green hydrogen production) are only starting to transition, but we expect they will occur rapidly as the other areas put more and more pressure on that sector to make similar adjustments. There has been good progress, in general, across many countries. However, it is very important to continue the efforts to find good pathways to net zero and make them a high priority.

14.2 Positive Transitions

With the transition to electric vehicles focused mainly on personal cars, one of the things to look at moving forward is the expansion of electric-based transportation beyond that category. In March and April 2024, the new emission standards for U.S. heavy-duty vehicles were introduced and published in the Federal Register (EPA, 2024). These new emission standards are for new vehicles in model year 2027, and there can be huge positive impacts from these new standards. The impacts are likely to be beneficial for both progress toward net zero and public health. Air quality has been negatively impacted by heavy-duty vehicles for over 50 years. The Clean Air Task Force estimates that in 2023, diesel vehicle emissions contributed to more than 8,000 deaths, 3,700 heart attacks, and about $100 billion in health-related losses in the U.S. (CATF, 2024). There has also been continued progress toward meeting other, generally

DOI: 10.1201/9781003396154-14

more stringent vehicle emission standards with zero-emission battery electric vehicles, plug-in hybrid electric vehicles, and hydrogen fuel cell vehicles. These new vehicles will be more efficient and contribute to the path to net zero.

There are efforts to reduce carbon dioxide emissions in the European Union through emission standards also (CATF, 2024). In fact, the European Union includes a number of countries that are world leaders in efforts to reduce greenhouse gas emissions. Two of the key goals the EU set in 2021 are to reduce net greenhouse gas emissions by 55% by 2030 and then to reach 100% (net zero) by 2050 (Sanderson and Wong, 2024). A proposed new intermediate target is to reduce net greenhouse gas emissions by 90% compared with 1990 levels by 2040.

Reducing emissions is one side of an equation, with the other side being carbon removal from the environment. From a net zero goal perspective, this is also an important part of the process to achieve that target (Sanderson and Wong, 2024). Carbon sequestration can occur by adding organic carbon to soils, removing carbon from seawater, removing carbon from the atmosphere, and land reclamation to improve soil health to enable productive agriculture and expanding forests. All of these can contribute to carbon removal from the ecosystem. The cost of these alternatives varies, of course, but the net costs also can depend on subsequent benefits as well. For instance, the associated benefits because adding organic carbon to soils (in the form of farmable land) may be sufficient to fully cover the expense of adding compost, manure, or biosolids. The establishment of additional forests to produce useful products may similarly have sufficient benefits to have a very low cost for carbon removal from the atmosphere.

There is a need to expand research and development on some of the more technology-based carbon removal processes, including carbon capture and storage and carbon removal from seawater. A positive development for a net zero transition would be to find several good ways to reduce the cost of carbon dioxide removal from the environment.

14.3 Recent Developments

Transitions to net zero emissions can take decades to complete, and one consequence of that long time frame is that the pace of those transitions is uncertain. It is possible, of course, that some forces will try to delay transitions, but it is also possible that there will be technological innovations, breakthroughs, or discoveries that accelerate transitions even more than our current projections. At this moment, there does seem to be something like a groundswell of interest, converging consensus, and resolutions to take action. This book is just one piece of that.

Another new publication in 2024 is a book on decarbonization strategies (Prasad et al., 2024), an approach that is along the same lines as the net zero goal but without a specific reference target. This book contains 27 chapters that include significant new content on wind energy, geothermal energy, hydrogen, electric vehicle developments, and many new aspects of carbon sequestration. For anyone interested in even more details on this aspect of reducing greenhouse gas emissions, this new book supplements the content and subjects covered here.

The U.S. National Academies of Sciences, Engineering, and Medicine also have their new report on accelerating decarbonization, which includes 13 chapters on many important topics related to pathways to net zero greenhouse gas emissions (NASEM, 2024). The report has about 80 recommendations related to economics, technology, policy, public participation, and other important issues. Chapter 5 of the National Academies report is on public engagement and addresses the importance of citizen participation in the transition to net zero. There is a need to have positive public participation, an educated public, and equity in the transition. Many decisions and actions involve owners of homes, vehicles, businesses, buildings, and land where education and understanding are needed to make good decisions. The public participates in voting for representatives and government officials who are expected to provide good leadership along pathways to net zero. This 2024 report is the second report by the National Academies related to decarbonization, with the first report published back in 2021 (NASEM, 2021). For anyone interested in even more details on these aspects of reducing greenhouse gas emissions (i.e., economics, politics, decision-making, and building involvement), these reports supplement the content and subjects covered here.

Many companies are including the goal of becoming a net zero company in their decisions to transition to renewable energy, net zero buildings, and electric vehicles. The Science Based Targets Initiative (SBTi) has published information to help corporations that want to be recognized as net zero corporations (SBTi, 2024). SBTi is working with other organizations to develop and publish standards that provide useful guidance for corporations that want to operate as a net zero organization.

14.4 Concluding Message

With a goal of reaching net zero by 2050, the clock is ticking. Far too much time has been wasted on distractions and digressions, and we now face a crisis-level need to take major and decisive actions. At the same time, though, there are reasons to be hopeful. The efforts to reach net zero are ongoing, and (as this book is being written, based on information available in June 2024) there are signs that the necessary transitions to make net zero a reality are finally happening. The need for new ideas and research progress continues to be important, the need for new investments and infrastructure is now obvious, and the need for policy and behavioral changes is being recognized as critical for success. It will be fascinating to see the next couple of decades as these different streams converge to make net zero emissions a reality. It will be fascinating to see the associated knock-on effects on health, geopolitical stability, and human welfare that can also happen if we do it right. For now, it is necessary to go forward by ending this book and getting to work.

References

CATF. 2024. EPA Releases Long-Awaited Greenhouse Gas Emission Standards for Heavy-Duty Vehicles. Clean Air Task Force. www.catf.us/

EPA. 2024. Final Rule: Greenhouse Gas Emissions Standards for Heavy-Duty Vehicles—Phase 3. Federal Register: 40 CFR Parts 86, 1036, 1037, 1039, 1054, 1065. www.epa.gov.

NASEM. 2021. Accelerating Decarbonization of the U.S. Energy System. The National Academies Press. https://doi.org/10.17226/25932.

NASEM. 2024. Accelerating Decarbonization in the United States: Technology, Policy, and Societal Dimensions. The National Academies Press. https://doi.org/10.17226/25931.

Prasad, M.N.V., Erickson, L.E., Nunes, F.C. and Ramadan, B.S., Eds. 2024. Decarbonization Strategies and Drivers to Achieve Carbon Neutrality for Sustainability. Elsevier, Amsterdam.

Sanderson, K. and Wong, C. 2024. EU unveils climate target: What scientists think. Nature 626: 467.

SBTi. 2024. SBTi Corporate Net-Zero Standard. Science Based Targets initiative, London, England. sciencebasedtargets.org.

15

A Moment of Truth

On June 5, 2024, the Secretary-General of the United Nations, Antonio Guterres, gave a special address on climate action titled "A Moment of Truth." He was speaking at the American Museum of Natural History, and below is a copy of his address based on the printed content provided by the United Nations (United Nations, 2024). We feel that this both summarizes many of the topics of this book as well as demonstrates the broad consensus in the responsible community about where we are (as of June 2024) and what needs to be done to make progress along the pathway to net zero.

> Dear friends of the planet,
> Today is World Environment Day.
> It is also the day that the European Commission's Copernicus Climate Change Service officially reports May 2024 as the hottest May in recorded history. This marks twelve straight months of the hottest months ever. For the past year, every turn of the calendar has turned up the heat. Our planet is trying to tell us something. But we don't seem to be listening.
>
> Dear Friends,
> The American Museum of Natural History is the ideal place to make the point. This great Museum tells the amazing story of our natural world. Of the vast forces that have shaped life on earth over billions of years. Humanity is just one small blip on the radar. But like the meteor that wiped out the dinosaurs, we're having an outsized impact.
>
> In the case of climate, we are not the dinosaurs. We are the meteor. We are not only in danger. We are the danger. But we are also the solution.
>
> So, dear friends, we are at a moment of truth.
> The truth is . . . almost ten years since the Paris Agreement was adopted, the target of limiting long-term global warming to 1.5 degrees Celsius is hanging by a thread.
>
> The truth is . . . the world is spewing emissions so fast that by 2030, a far higher temperature rise would be all but guaranteed. Brand new data from leading climate scientists released today show the remaining carbon

DOI: 10.1201/9781003396154-15

budget to limit long-term warming to 1.5 degrees is now around 200 billion tonnes. That is the maximum amount of carbon dioxide that the earth's atmosphere can take if we are to have a fighting chance of staying within the limit.

The truth is . . . we are burning through the budget at reckless speed—spewing out around 40 billion tonnes of carbon dioxide a year. We can all do the math.

At this rate, the entire carbon budget will be busted before 2030.

The truth is . . . global emissions need to fall nine per cent every year until 2030 to keep the 1.5 degree limit alive. But they are heading in the wrong direction. Last year they rose by one per cent.

The truth is . . . we already face incursions into 1.5-degree territory.

The World Meteorological Organisation reports today that there is an eighty per cent chance the global annual average temperature will exceed the 1.5 degree limit in at least one of the next five years. In 2015, the chance of such a breach was near zero.

And there's a fifty-fifty chance that the average temperature for the entire next five-year period will be 1.5 degrees higher than pre-industrial times.

We are playing Russian roulette with our planet. We need an exit ramp off the highway to climate hell. And the truth is . . . we have control of the wheel. The 1.5 degree limit is still just about possible. Let's remember—it's a limit for the long-term—measured over decades, not months or years. So, stepping over the threshold 1.5 for a short time does not mean the long-term goal is shot. It means we need to fight harder.

Now.

The truth is . . . the battle for 1.5 degrees will be won or lost in the 2020s—under the watch of leaders today. All depends on the decisions those leaders take—or fail to take—especially in the next eighteen months. It's climate crunch time. The need for action is unprecedented but so is the opportunity—not just to deliver on climate, but on economic prosperity and sustainable development. Climate action cannot be captive to geo-political divisions.

So, as the world meets in Bonn for climate talks, and gears up for the G7 and G20 Summits, the United Nations General Assembly, and COP29, we need maximum ambition, maximum acceleration, maximum cooperation—in a word maximum action.

So dear friends,

Why all this fuss about 1.5 degrees? Because our planet is a mass of complex, connected systems. And every fraction of a degree of global heating counts. The difference between 1.5 and two degrees could be the difference between extinction and survival for some small island states and coastal communities. The difference between minimizing climate chaos or crossing dangerous tipping points. 1.5 degrees is not a target. It is not a goal. It is a physical limit. Scientists have alerted us that temperatures rising higher would likely mean:

The collapse of the Greenland Ice Sheet and the West Antarctic Ice Sheet with catastrophic sea level rise; The destruction of tropical coral reef systems and the livelihoods of 300 million people; The collapse of the Labrador Sea Current that would further disrupt weather patterns in Europe; And widespread permafrost melt that would release devastating levels of methane, one of the most potent heat-trapping gasses.

Even today, we're pushing planetary boundaries to the brink—shattering global temperature records and reaping the whirlwind.

And it is a travesty of climate justice that those least responsible for the crisis are hardest hit: the poorest people; the most vulnerable countries; Indigenous Peoples; women and girls. The richest one per cent emit as much as two-thirds of humanity. And extreme events turbocharged by climate chaos are piling up: Destroying lives, pummelling economies, and hammering health; Wrecking sustainable development; forcing people from their homes; and rocking the foundations of peace and security—as people are displaced and vital resources depleted.

Already this year, a brutal heatwave has baked Asia with record temperatures—shrivelling crops, closing schools, and killing people. Cities from New Delhi, to Bamako, to Mexico City are scorching. Here in the US, savage storms have destroyed communities and lives. We've seen drought disasters declared across southern Africa; Extreme rains flood the Arabian Peninsula, East Africa and Brazil; And a mass global coral bleaching caused by unprecedented ocean temperatures, soaring past the worst predictions of scientists.

The cost of all this chaos is hitting people where it hurts: From supply-chains severed, to rising prices, mounting food insecurity, and uninsurable homes and businesses. That bill will keep growing. Even if emissions hit zero tomorrow, a recent study found that climate chaos will still cost at least $38 trillion a year by 2050. Climate change is the mother of all stealth taxes paid by everyday people and vulnerable countries and communities. Meanwhile, the Godfathers of climate chaos—the fossil fuel industry—rake in record profits and feast off trillions in taxpayer-funded subsidies.

Dear friends,
We have what we need to save ourselves. Our forests, our wetlands, and our oceans absorb carbon from the atmosphere. They are vital to keeping 1.5 alive, or pulling us back if we do overshoot that limit. We must protect them. And we have the technologies we need to slash emissions. Renewables are booming as costs plummet and governments realise the benefits of cleaner air, good jobs, energy security, and increased access to power. Onshore wind and solar are the cheapest source of new electricity in most of the world—and have been for years. Renewables already make up thirty percent of the world's electricity supply. And clean energy investments reached a record high last year—almost doubling in the last ten [years]. Wind and solar are now growing

faster than any electricity source in history. Economic logic makes the end of the fossil fuel age inevitable.

The only questions are: Will that end come in time? And will the transition be just?

Dear friends,

We must ensure the answer to both questions is: yes. And we must secure the safest possible future for people and planet. That means taking urgent action, particularly over the next eighteen months: To slash emissions; To protect people and nature from climate extremes;

To boost climate finance; And to clamp down on the fossil fuel industry. Let me take each element in turn.

First, huge cuts in emissions. Led by the huge emitters. The G20 countries produce eighty percent of global emissions—they have the responsibility, and the capacity, to be out in front. Advanced G20 economies should go furthest, fastest; And show climate solidarity by providing technological and financial support to emerging G20 economies and other developing countries. Next year, governments must submit so-called nationally determined contributions—in other words, national climate action plans. And these will determine emissions for the coming years.

At COP28, countries agreed to align those plans with the 1.5 degree limit. These national plans must include absolute emission reduction targets for 2030 and 2035. They must cover all sectors, all greenhouse gases, and the whole economy. And they must show how countries will contribute to the global transitions essential to 1.5 degrees—putting us on a path to global net zero by 2050; to phase out fossil fuels; and to hit global milestones along the way, year after year; and decade after decade. That includes, by 2030, contributing to cutting global production and consumption of all fossil fuels by at least thirty percent; and making good on commitments made at COP28—on ending deforestation, doubling energy efficiency and tripling renewables.

Every country must deliver and play their rightful part. That means that G20 leaders working in solidarity to accelerate a just global energy transition aligned with the 1.5 degree limit. They must assume their responsibilities. We need cooperation, not finger-pointing. It means the G20 aligning their national climate action plans, their energy strategies, and their plans for fossil fuel production and consumption, within a 1.5 degree future. It means the G20 pledging to reallocate subsidies from fossil fuels to renewables, storage, and grid modernisation, and support for vulnerable communities. It means the G7 and other OECD countries committing: to end coal by 2030; and to create fossil-fuel free power systems, and reduce oil and gas supply and demand by sixty percent—by 2035. It means all countries ending new coal projects—now. Particularly in Asia, home to ninety-five percent of planned new coal power capacity. It means non-OECD countries creating climate action plans to put them on a path to ending coal power by 2040.

And it means developing countries creating national climate action plans that double as investment plans, spurring sustainable development, and meeting soaring energy demand with renewables.

The United Nations is mobilizing our entire system to help developing countries to achieve this through our Climate Promise initiative. Every city, region, industry, financial institution, and company must also be part of the solution. They must present robust transition plans by COP30 next year in Brazil—at the latest: Plans aligned with 1.5 degrees, and the recommendations of the UN High-Level Expert Group on Net Zero. Plans that cover emissions across the entire value chain; That include interim targets and transparent verification processes; And that steer clear of the dubious carbon offsets that erode public trust while doing little or nothing to help the climate.

We can't fool nature. False solutions will backfire. We need high integrity carbon markets that are credible and with rules consistent with limiting warming to 1.5 degrees. I also encourage scientists and engineers to focus urgently on carbon dioxide removal and storage—to deal safely and sustainably with final emissions from the heavy industries hardest to clean. And I urge governments to support them. But let me be clear: These technologies are not a silver bullet; they cannot be a substitute for drastic emissions cuts or an excuse to delay fossil fuel phase-out.

But we need to act on every front.

Dear Friends,
The second area for action is ramping up protection from the climate chaos of today and tomorrow. It is a disgrace that the most vulnerable are being left stranded, struggling desperately to deal with a climate crisis they did nothing to create. We cannot accept a future where the rich are protected in air-conditioned bubbles, while the rest of humanity is lashed by lethal weather in unliveable lands. We must safeguard people and economies. Every person on Earth must be protected by an early warning system by 2027. I urge all partners to boost support for the United Nations Early Warnings for All action plan. In April, the G7 launched the Adaptation Accelerator Hub. By COP29, this initiative must be translated into concrete action—to support developing countries in creating adaptation investment plans, and putting them into practice. And I urge all countries to set out their adaptation and investment needs clearly in their new national climate plans. But change on the ground depends on money on the table.

For every dollar needed to adapt to extreme weather, only about five cents is available.

As a first step, all developed countries must honour their commitment to double adaptation finance to at least $40 billion a year by 2025. And they must set out a clear plan to close the adaptation finance gap by COP29 in November. But we also need more fundamental reform. That leads me onto my third point: finance.

Dear friends,
If money makes the world go round, today's unequal financial flows are sending us spinning towards disaster. The global financial system must be part of the climate solution. Eye-watering debt repayments are drying up funds for climate action. Extortion-level capital costs are putting renewables virtually out of reach for most developing and emerging economies. Astoundingly—and despite the renewables boom of recent years—clean energy investments in developing and emerging economies outside of China have been stuck at the same levels since 2015. Last year, just fifteen per cent of new clean energy investment went to emerging markets and developing economies outside China—countries representing nearly two-thirds of the world's population.

And Africa was home to less than one percent of last year's renewables installations, despite its wealth of natural resources and vast renewables potential.

The International Energy Agency reports that clean energy investments in developing and emerging economies beyond China need to reach up to $1.7 trillion a year by the early 2030s. In short, we need a massive expansion of affordable public and private finance to fuel ambitious new climate plans and deliver clean, affordable energy for all. This September's Summit of the Future is an opportunity to push reform of the international financial architecture and action on debt. I urge countries to take it. And I urge the G7 and G20 Summits to commit to using their influence within Multilateral Development Banks to make them better, bigger, and bolder. And able to leverage far more private finance at reasonable cost. Countries must make significant contributions to the new Loss and Damage Fund. And ensure that it is open for business by COP29. And they must come together to secure a strong finance outcome from COP this year—one that builds trust and confidence, catalyses the trillions needed, and generates momentum for reform of the international financial architecture.

But none of this will be enough without new, innovative sources of funds. It is [high] time to put an effective price on carbon and tax the windfall profits of fossil fuel companies. By COP29, we need early movers to go from exploring to implementing solidarity levies on sectors such as shipping, aviation, and fossil fuel extraction—to help fund climate action. These should be scalable, fair, and easy to collect and administer. None of this is charity. It is enlightened self-interest. Climate finance is not a favour. It is fundamental element to a liveable future for all.

Dear friends,
Fourth and finally, we must directly confront those in the fossil fuel industry who have shown relentless zeal for obstructing progress—over decades. Billions of dollars have been thrown at distorting the truth, deceiving the public, and sowing doubt. I thank the academics and the activists, the journalists and the whistleblowers, who have exposed those tactics—often at great personal

and professional risk. I call on leaders in the fossil fuel industry to understand that if you are not in the fast lane to clean energy transformation, you are driving your business into a dead end—and taking us all with you. Last year, the oil and gas industry invested a measly 2.5 percent of its total capital spending on clean energy. Doubling down on fossil fuels in the twenty-first century, is like doubling down on horse-shoes and carriage-wheels in the nineteenth.

So, to fossil fuel executives, I say: your massive profits give you the chance to lead the energy transition. Don't miss it.

Financial institutions are also critical because money talks. It must be a voice for change.

I urge financial institutions to stop bankrolling fossil fuel destruction and start investing in a global renewables revolution; To present public, credible and detailed plans to transition [funding] from fossil fuels to clean energy with clear targets for 2025 and 2030; And to disclose your climate risks—both physical and transitional—to your shareholders and regulators. Ultimately such disclosure should be mandatory.

Dear friends,

Many in the fossil fuel industry have shamelessly greenwashed, even as they have sought to delay climate action—with lobbying, legal threats, and massive ad campaigns. They have been aided and abetted by advertising and PR companies—Mad Men—remember the TV series—fuelling the madness. I call on these companies to stop acting as enablers to planetary destruction. Stop taking on new fossil fuel clients, from today, and set out plans to drop your existing ones. Fossil fuels are not only poisoning our planet—they're toxic for your brand.

Your sector is full of creative minds who are already mobilising around this cause. They are gravitating towards companies that are fighting for our planet—not trashing it.

I also call on countries to act. Many governments restrict or prohibit advertising for products that harm human health—like tobacco. Some are now doing the same with fossil fuels. I urge every country to ban advertising from fossil fuel companies. And I urge news media and tech companies to stop taking fossil fuel advertising. We must all deal also with the demand side. All of us can make a difference, by embracing clean technologies, phasing down fossil fuels in our own lives, and using our power as citizens to push for systemic change. In the fight for a liveable future, people everywhere are far ahead of politicians. Make your voices heard and your choices count.

Dear friends,

We do have a choice. Creating tipping points for climate progress—or careening to tipping points for climate disaster. No country can solve the climate crisis in isolation. This is an all-in moment. The United Nations is all-in—working to build trust, find solutions, and inspire the cooperation our world so desperately needs. And to young people, to civil society, to cities, regions, businesses and others who have been leading the charge towards a safer, cleaner world, I say: Thank you.

You are on the right side of history. You speak for the majority. Keep it up. Don't lose courage. Don't lose hope.

It is we the Peoples versus the polluters and the profiteers. Together, we can win. But it's time for leaders to decide whose side they're on. Tomorrow it will be too late.Now is the time to mobilise, now is the time to act, now is the time to deliver.

This is our moment of truth.

And I thank you.

Reference

United Nations. 2024. Secretary-General's Special Address on Climate Action "A Moment of Truth". United Nations, New York City. https://www.un.org/sg/en/content/sg/speeches/2024-06-05/secretary-generals-special-address-climate-action-moment-of-truth%C2%A0

Index

Note: Page numbers in **bold** indicate a table on the corresponding page.

Printed in the United States
by Baker & Taylor Publisher Services